Praise for Brian Chr...

THE MOST HUMAN HUMAN

"An irreverent picaresque. . . . What Christian learns along the way is that if machines win the imitation game as often as they do, it's not because they're getting better at acting human, it's because we're getting worse. . . . As *The Most Human Human* demonstrates, Christian has taken his own words to heart. An authentic son of [Robert] Frost, he learns by going where he has to go, and in doing so proves that both he and his book deserve their title."

—*The New York Times Book Review*

"Christian covers a great deal of ground with admirable clarity but with a lightness of touch. . . . He also has a real knack for summing up key ideas by applying them to real-life situations. . . . Did Christian become 'The Most Human Human'? You'll have to read the book to find out." —*The Wall Street Journal*

"Excellent." —NPR's *Radiolab*

"Incredibly engrossing." —*The Onion* A. V. Club

"Entertaining and informative." —*The Economist*

"Exhilarating. . . . Reading it, I constantly found my mind pinging off of whatever Christian was discussing and into flights of exploratory speculation about the amount of information encoded in the seemingly routine exchanges of small talk or the reasons why it's much harder to tell a false story in reverse chronological order. It's an unusual book whose primary gift lies in distracting you from itself. I'd like to see the computers come up with something like that."

—Laura Miller, *Salon*

"A charming, friendly, and often funny read." —*The Boston Globe*

"Immensely ambitious and bold, intellectually provocative, while at the same time entertaining and witty—a delightful book about how to live a meaningful, thriving life."
—Alan Lightman, author of *Einstein's Dreams* and *Ghost*

"A book exploring the wild frontiers of chat-bots is appealing enough; I never expected to discover in its pages such an eye-opening inquest into human imagination, thought, conversation, love and deception. Who would have guessed that the best way to understand humanity was to study its imitators?"
—David Eagleman, author of *Sum* and *Incognito*

"Remarkable, enjoyable, heartening. A philosophical joyride connecting the thoughts of Aristotle with David Brent. . . . The day that a machine creates work of such wit and originality, we should all be very worried." —*The Times* (London)

"This is a strange, fertile, and sometimes beautiful book. Brian Christian writes with a rare combination of what Pascal took to be two contrary mind-sets: the spirit of geometry and the spirit of finesse. He takes both the deep limitations and halting progress of artificial intelligence as an occasion for thinking about the most human activity—the art of conversation."
—Matthew B. Crawford, author of *Shop Class as Soulcraft*

"Lively and thought-stirring. . . . An invaluable sourcebook on computing in modern-day life." —*New Statesman*

"Fast-paced, witty, and thoroughly winning."
—*Publishers Weekly* (starred review)

Brian Christian

THE MOST HUMAN HUMAN

Brian Christian's work has appeared in *The Atlantic, Wired, The Wall Street Journal,* and many literary and scientific publications. He has been featured on *The Daily Show with Jon Stewart,* NPR's *Radiolab,* and *The Charlie Rose Show,* and has lectured at Google, Microsoft, the London School of Economics, and elsewhere. An award-winning poet, Christian holds a degree in philosophy and computer science from Brown University and an MFA in poetry from the University of Washington. *The Most Human Human,* a *Wall Street Journal* bestseller, has been translated into nine languages. Christian lives in Philadelphia.

www.brchristian.com

THE
MOST
HUMAN
HUMAN

What Artificial Intelligence Teaches Us

About Being Alive

Brian Christian

ANCHOR BOOKS

A Division of Penguin Random House LLC

New York

FIRST ANCHOR BOOKS EDITION, FEBRUARY 2012

Copyright © 2011 by Brian Christian

All rights reserved. Published in the United States by Anchor Books,
a division of Penguin Random House LLC, New York, and distributed in
Canada by Random House of Canada, a division of Penguin Random House
Canada Limited, Toronto. Originally published in hardcover as *The Most
Human Human: What Talking with Computers Teaches Us About What It
Means to Be Alive* in the United States by Doubleday, a division of
Penguin Random House LLC, New York, in 2011.

Anchor Books and colophon are registered trademarks of
Penguin Random House LLC.

Portions of this work were previously published in *The Atlantic*.

Grateful acknowledgment is made to Richard Wilbur for permission to reprint
a portion of "The Beautiful Changes."

The Library of Congress has cataloged the Doubleday edition as follows:
Christian, Brian, 1984–
The most human human / Brian Christian.
p. cm.
1. Philosophical anthropology.   2. Human beings.   3. Turing test.   I. Title.
BD450.C5356 2011
128—dc22
2010048572

Anchor Books Trade Paperback ISBN: 978-0-307-47670-8
eBook ISBN: 978-0-385-53307-2

Book design by Michael Collica

www.anchorbooks.com

Printed in the United States of America
10 9 8 7 6

Today's person spends way more time in front of screens. In fluorescent-lit rooms, in cubicles, being on one end or the other of an electronic data transfer. And what is it to be human and alive and exercise your humanity in that kind of exchange?

—DAVID FOSTER WALLACE

For my teachers

Contents

The beautiful changes as a forest is changed
By a chameleon's tuning his skin to it;
As a mantis, arranged
On a green leaf, grows
Into it, makes the leaf leafier . . .

— RICHARD WILBUR

I think metaphysics is good if it improves
everyday life; otherwise forget it.

— ROBERT PIRSIG

As President, I believe that robotics can
inspire young people to pursue science and
engineering. And I also want to keep an eye
on those robots, in case they try anything.

— BARACK OBAMA

THE
MOST
HUMAN
HUMAN

0. Prologue

Claude Shannon, artificial intelligence pioneer and founder of information theory, met his wife, Mary Elizabeth, at work. This was Bell Labs in Murray Hill, New Jersey, the early 1940s. He was an engineer, working on wartime cryptography and signal transmission.

She was a computer.

1. Introduction:
The Most Human Human

I wake up five thousand miles from home in a hotel room with no shower: for the first time in fifteen years, I take a bath. I eat, as is traditional, some slightly ominous-looking tomatoes, some baked beans, and four halves of white toast that come on a tiny metal rack, shelved vertically, like books. Then I step out into the salty air and walk the coastline of the country that invented my language, despite my not being able to understand a good portion of the signs I pass on my way—LET AGREED, one says, prominently, in large print, and it means nothing to me.

I pause, and stare dumbly at the sea for a moment, parsing and reparsing the sign in my head. Normally these kinds of linguistic curiosities and cultural gaps interest and intrigue me; today, though, they are mostly a cause for concern. In the next two hours I will sit down at a computer and have a series of five-minute instant-message chats with several strangers. At the other end of these chats will be a psychologist, a linguist, a computer scientist, and the host of a popular British technology show. Together they form a judging panel, and my goal in these conversations is one of the strangest things I've ever been asked to do.

I must convince them that I'm human.

Fortunately, I *am* human; unfortunately, it's not clear how much that will help.

The Turing Test

Each year, the artificial intelligence (AI) community convenes for the field's most anticipated and controversial annual event—a competition called the Turing test. The test is named for British mathematician Alan Turing, one of the founders of computer science, who in 1950 attempted to answer one of the field's earliest questions: *Can machines think?* That is, would it ever be possible to construct a computer so sophisticated that it could actually be said to be thinking, to be intelligent, to have a mind? And if indeed there were, someday, such a machine: How would we know?

Instead of debating this question on purely theoretical grounds, Turing proposed an experiment. A panel of judges poses questions by computer terminal to a pair of unseen correspondents, one a human "confederate," the other a computer program, and attempts to discern which is which. There are no restrictions on what can be said: the dialogue can range from small talk to the facts of the world (e.g., how many legs ants have, what country Paris is in) to celebrity gossip and heavy-duty philosophy—the whole gamut of human conversation. Turing predicted that by the year 2000, computers would be able to fool 30 percent of human judges after five minutes of conversation, and that as a result "one will be able to speak of machines thinking without expecting to be contradicted."

Turing's prediction has not come to pass; at the 2008 contest, however, held in Reading, England, the top program came up shy of that mark by just a single vote. The 2009 test in Brighton could be the decisive one.

And I am participating in it, as one of four human confederates going head-to-head (head-to-motherboard?) against the top AI programs. In each of several rounds, I, along with the other confederates, will be paired off with an AI program and a judge—and will have the task of convincing the latter that I am, in fact, human.

The judge will talk to one of us for five minutes, then the other, and then has ten minutes to reflect and make his choice about which one of us he believes is the human. Judges will also note, on a sliding scale, their confidence in this judgment—this is used in part as a tie-breaking measure. The program that receives the highest share of votes and confidence from the judges each year (regardless of whether it "passes the Turing test" by fooling 30 percent of them) is awarded the "Most Human Computer" title. It is this title that the research teams are all gunning for, the one that the money awards, the one with which the organizers and spectators are principally concerned. But there is also, intriguingly, another title, one given to the *confederate* who elicited the greatest number of votes and greatest confidence from the judges: the "Most Human Human" award.

One of the first winners, in 1994, was *Wired* columnist Charles Platt. How'd he do it? By "being moody, irritable, and obnoxious," he says—which strikes me as not only hilarious and bleak but also, in some deeper sense, a call to arms: How, in fact, do we be the most human humans we can be—not only under the constraints of the test, but in life?

Joining the Confederacy

The sponsor and organizer of the Turing test (this particular incarnation of which is known as the Loebner Prize) is a colorful and somewhat curious figure: plastic roll-up portable disco dance floor baron Hugh Loebner. When asked his motives for backing and orchestrating this annual Turing test, Loebner cites *laziness,* of all things: his utopian future, apparently, is one in which unemployment rates are nearly 100 percent and virtually all of human endeavor and industry is outsourced to intelligent machines. I must say, this vision of the future makes me feel little but despair, and I have my own, quite different ideas about what an AI-populated world would look like and reasons for participating in the test. But in any event, the central question of how computers are reshaping our sense of self, and what the ramifications of that process will be, is clearly the crucial one.

Not entirely sure how to go about becoming a confederate, I started at the top: by trying to reach Hugh Loebner himself. I quickly found his website, where, amid a fairly inscrutable amalgam of material about crowd-control stanchions,[1] sex-work activism,[2] and a scandal involving the composition of Olympic medals,[3] I was able to find information on his eponymous prize, along with his email address. He replied by giving me the name of Philip Jackson, a professor at the University of Surrey, who is the one in charge of the logistics for this year's Loebner Prize contest in Brighton, where it will be held under the auspices of the 2009 Interspeech conference on speech and communication science.

I was able to get in touch via Skype with Professor Jackson, a young, smart guy with the distinct brand of harried enthusiasm that characterizes an overworked but fresh-faced academic. That and his charming Briticisms, like pronouncing "skeletal" so it'd rhyme with "a beetle": I liked him immediately.

He asked me about myself, and I explained that I'm a nonfiction writer of science and philosophy, specifically of the ways in which science and philosophy intersect with daily life, and that I'm fascinated by the idea of the Turing test and of the "Most Human Human." For one, there's a romantic notion as a confederate of *defending the human race,* à la Garry Kasparov vs. Deep Blue—and soon, Ken

---

1. Crowd-control stanchions seem to have recently replaced portable disco dance floors as the flagship product of Loebner's company, Crown Industries, which is the Loebner Prize's chief sponsor.
2. Surely I'm not the only one who finds it ironic that a man who's committed himself to advancing the progress of interaction with *artificial* entities has resigned himself—as he has discussed openly in the pages of the *New York Times* and on several television talk shows—to paying, whether happily or unhappily, for *human* intimacy?
3. Apparently the "gold" medals are actually silver medals *dipped in gold*—which is, admittedly, a bit bizarre, although it seems to have caused Loebner more than a decade of outrage, which over the years has vented itself in the form of picketing, speeches, and a newsletter called *Pants on Fire News.*

Jennings of *Jeopardy!* fame vs. the latest IBM system, Watson. (The mind also leaps to other, more *Terminator–* and *The Matrix–*type fantasies, although the Turing test promises to involve *significantly* fewer machine guns.) When I read that the machines came up shy of passing the 2008 test by just one single vote, and realized that 2009 might be the year they finally cross the threshold, a steely voice inside me rose up seemingly out of nowhere. *Not on my watch.*

More than this, though, the test raises a number of questions, exciting as well as troubling, at the intersection of computer science, cognitive science, philosophy, and daily life. As someone who has studied and written about each of these areas, and who has published peer-reviewed cognitive science research, I find the Turing test particularly compelling for the way it manages to draw from and connect them all. As we chatted, I told Professor Jackson that I thought I might have something rather unique to bring to the Loebner Prize, in terms of both the actual performance of being a confederate and relating that experience, along with the broader questions and issues raised by the test, to a large audience—which would start what I think could be a fascinating and important conversation in the public culture at large. It wasn't hard to get him to agree, and soon my name was on the confederate roster.

After briefing me a bit on the logistics of the competition, he gave me the advice I had heard from confederates past to expect: "There's not much more you need to know, really. You *are* human, so just be yourself."

"Just be yourself"—this has been, in effect, the confederate motto since the first Loebner Prize in 1991, but seems to me like a somewhat naive overconfidence in human instincts—or at worst, fixing the fight. The AI programs we go up against are often the result of decades of work—then again, so are we. But the AI research teams have huge databases of test runs of their programs, and they've done statistical analysis on these archives: they know how to deftly guide the conversation away from their shortcomings and toward their strengths, what conversational routes lead to deep exchange and which ones

fizzle—the average confederate off the street's instincts aren't likely to be so good. This is a strange and deeply interesting point, of which the perennial demand in our society for conversation, public speaking, and dating coaches is ample proof. The transcripts from the 2008 contest show the judges being downright apologetic to the human confederates that they can't make better conversation—"i feel sorry for the [confederates], i reckon they must be getting a bit bored talking about the weather," one says, and another offers, meekly, "sorry for being so banal"—meanwhile, the computer in the other window is apparently charming the pants off the judge, who in no time at all is gushing lol's and :P's. We can do better.

So, I must say, my intention from the start was to be as thoroughly disobedient to the organizers' advice to "just show up at Brighton in September and 'be myself' " as possible—spending the months leading up to the test gathering as much information, preparation, and experience as possible and coming to Brighton ready to give it everything I had.

Ordinarily, there wouldn't be very much odd about this notion at all, of course—we train and prepare for tennis competitions, spelling bees, standardized tests, and the like. But given that the Turing test is meant to evaluate *how human* I am, the implication seems to be that being human (and being oneself) is about more than simply showing up. I contend that it is. What exactly that "more" entails will be a main focus of this book—and the answers found along the way will be applicable to a lot more in life than just the Turing test.

Falling for Ivana

A rather strange, and more than slightly ironic, cautionary tale: Dr. Robert Epstein, UCSD psychologist, editor of the scientific volume *Parsing the Turing Test,* and co-founder, with Hugh Loebner, of the Loebner Prize, subscribed to an online dating service in the winter of 2007. He began writing long letters to a Russian woman named

Ivana, who would respond with long letters of her own, describing her family, her daily life, and her growing feelings for Epstein. Eventually, though, something didn't feel right; long story short, Epstein ultimately realized that he'd been exchanging lengthy love letters for *over four months* with—you guessed it—a computer program. Poor guy: it wasn't enough that web-ruffians spam his email box every day, now they have to spam his heart?

On the one hand, I want to simply sit back and laugh at the guy—he *founded* the Loebner Prize, for Christ's sake! What a chump! Then again, I'm also sympathetic: the unavoidable presence of spam in the twenty-first century not only clogs the inboxes and bandwidth of the world (roughly 97 percent of *all email messages* are spam—we are talking tens of billions a day; you could *literally* power a small nation[4] with the amount of electricity it takes to process the world's daily spam), but does something arguably worse—it erodes our sense of trust. I hate that when I get messages from my friends I have to expend at least a modicum of energy, at least for the first few sentences, deciding whether it's really *them* writing. We go through digital life, in the twenty-first century, with our guards up. All communication is a Turing test. All communication is suspect.

That's the pessimistic version, and here's the optimistic one. I'll bet that Epstein learned a lesson, and I'll bet that lesson was a lot more complicated and subtle than "trying to start an online relationship with someone from Nizhny Novgorod was a dumb idea." I'd like to think, at least, that he's going to have a lot of thinking to do about why it took him four months to realize that there was no actual exchange occurring between him and "Ivana," and that in the future he'll be quicker to the real-human-exchange draw. And that his *next* girlfriend, who hopefully not only is a bona fide *Homo sapiens* but also lives fewer than eleven time zones away, may have "Ivana," in a weird way, to thank.

---

4. Say, Ireland.

The Illegitimacy of the Figurative

When Claude Shannon met Betty at Bell Labs in the 1940s, she was indeed a computer. If this sounds odd to us in any way, it's worth knowing that nothing at all seemed odd about it to them. Nor to their co-workers: to their Bell Labs colleagues their romance was a perfectly normal one, typical even. Engineers and computers wooed all the time.

It was Alan Turing's 1950 paper "Computing Machinery and Intelligence" that launched the field of AI as we know it and ignited the conversation and controversy over the Turing test (or the "Imitation Game," as Turing initially called it) that has continued to this day—but modern "computers" are nothing like the "computers" of Turing's time. In the early twentieth century, before a "computer" was one of the digital processing devices that so proliferate in our twenty-first-century lives—in our offices, in our homes, in our cars, and, increasingly, in our pockets—it was something else: a job description.

From the mid-eighteenth century onward, computers, frequently women, were on the payrolls of corporations, engineering firms, and universities, performing calculations and doing numerical analysis, sometimes with the use of a rudimentary calculator. These original, human computers were behind the calculations for everything from the first accurate predictions for the return of Halley's comet—early proof of Newton's theory of gravity, which had only been checked against planetary orbits before—to the Manhattan Project, where Nobel laureate physicist Richard Feynman oversaw a group of human computers at Los Alamos.

It's amazing to look back at some of the earliest papers in computer science, to see the authors attempting to explain, for the first time, what exactly these new contraptions were. Turing's paper, for instance, describes the unheard-of "digital computer" by making analogies to a *human* computer: "The idea behind digital computers may be explained by saying that these machines are intended to carry out any operations which could be done by a human computer." Of course in the decades to come we know that the quotation marks migrated, and now it is the digital computer that is not only the default term, but the *literal* one. And it is the *human* "computer" that is relegated to the illegitimacy of the figurative. In the mid-twentieth century, a piece of cutting-edge mathematical gadgetry was "like a computer." In the twenty-first century, it is the *human* math whiz that is "like a computer." An odd twist: we're *like* the thing that used to be *like* us. We imitate our old imitators, one of the strange reversals of fortune in the long saga of human uniqueness.

The Sentence

Harvard psychologist Daniel Gilbert says that every psychologist must, at some point in his or her career, write a version of "The Sentence." Specifically, The Sentence reads like this: "The human being is the only animal that _____." Indeed, it seems that philosophers, psychologists, and scientists have been writing and rewriting this sentence since the beginning of recorded history. The story of humans' sense of self is, you might say, the story of failed, debunked versions of The Sentence. Except now it's not just the animals that we're worried about.

We once thought humans were unique for having a language with syntactical rules, but this isn't so;[5] we once thought humans were

---

5. Michael Gazzaniga, in *Human,* quotes Great Ape Trust primatologist Sue Savage-Rumbaugh: "First the linguists said we had to get our animals to use signs in a symbolic way if we wanted to say they learned language. OK, we did that, and then they said, 'No, that's not language, because you don't have syntax.' So we proved our apes could produce some combinations of signs, but the linguists said that wasn't enough syntax, or the right syntax. They'll never agree that we've done enough."

unique for using tools, but this isn't so;[6] we once thought humans were unique for being able to do mathematics, and now we can barely imagine being able to do what our calculators can.

There are several components to charting the evolution of The Sentence. One is a historical look at how various developments—in our knowledge of the world as well as our technical capabilities—have altered its formulations over time. From there, we can look at how these different theories have shaped humankind's sense of its own identity. For instance, are artists more valuable to us than they were before we discovered how difficult art is for computers?

Last, we might ask ourselves: Is it appropriate to allow our definition of our own uniqueness to be, in some sense, *reactionary* to the advancing front of technology? And why is it that we are so compelled to feel unique in the first place?

"Sometimes it seems," says Douglas Hofstadter, "as though each new step towards AI, rather than producing something which everyone agrees is real intelligence, merely reveals what real intelligence is *not*." While at first this seems a consoling position—one that keeps our unique claim to thought intact—it does bear the uncomfortable appearance of a gradual retreat, the mental image being that of a medieval army withdrawing from the castle to the keep. But the retreat can't continue indefinitely. Consider: if *everything* of which we regarded "thinking" to be a hallmark turns out not to involve it, then . . . what is thinking? It would seem to reduce to either an

---

6. Octopuses, for instance, were discovered in 2009 to use coconut shells as "body armor." The abstract of the paper that broke the news tells the story of our ever-eroding claim to uniqueness: "Originally regarded as a defining feature of our species, tool-use behaviours have subsequently been revealed in other primates and a growing spectrum of mammals and birds. Among invertebrates, however, the acquisition of items that are deployed later has not previously been reported. We repeatedly observed soft-sediment dwelling octopuses carrying around coconut shell halves, assembling them as a shelter only when needed."

epiphenomenon—a kind of "exhaust" thrown off by the brain—or, worse, an illusion.

Where is the keep of our *selfhood*?

The story of the twenty-first century will be, in part, the story of the drawing and redrawing of these battle lines, the story of *Homo sapiens* trying to stake a claim on shifting ground, flanked on both sides by beast and machine, pinned between meat and math.

And here's a crucial, related question: Is this retreat a good thing or a bad thing? For instance, does the fact that computers are so good at mathematics in some sense *take away* an arena of human activity, or does it *free* us from having to do a nonhuman activity, liberating us into a more human life? The latter view would seem to be the more appealing, but it starts to seem less so if we can imagine a point in the future where the number of "human activities" left to be "liberated" into has grown uncomfortably small. What then?

Inverting the Turing Test

There are no broader philosophical implications . . .
It doesn't connect to or illuminate anything.

—NOAM CHOMSKY, IN AN EMAIL TO THE AUTHOR

Alan Turing proposed his test as a way to measure the progress of technology, but it just as easily presents us a way to measure our *own*. Oxford philosopher John Lucas says, for instance, that if we fail to prevent the machines from passing the Turing test, it will be "not because machines are so intelligent, but because humans, many of them at least, are so wooden."

Here's the thing: beyond its use as a technological benchmark, beyond even the philosophical, biological, and moral questions it poses, the Turing test is, at bottom, about the act of communication. I see its deepest questions as practical ones: How do we connect meaningfully with each other, as meaningfully as possible, within the limits of language and time? How does empathy work? What is the process by which someone comes into our life and comes to mean something to us? These, to me, are the test's most central questions—the most central questions of being human.

Part of what's fascinating about studying the programs that have done well at the Turing test is that it is a (frankly, sobering) study of how conversation can work in the total absence of emotional intimacy. A look at the transcripts of Turing tests past is in some sense a tour of the various ways in which we demur, dodge the question, lighten the mood, change the subject, distract, burn time: what shouldn't pass as real conversation at the Turing test probably shouldn't be allowed to pass as real human conversation, either.

There are a number of books written about the technical side of the Turing test: for instance, how to cleverly design Turing test programs—called chatterbots, chatbots, or just bots. In fact, almost everything written at a practical level about the Turing test is about how to make good bots, with a small remaining fraction about how to be a good judge. But nowhere do you read how to be a good confederate. I find this odd, since the confederate side, it seems to me, is where the stakes are highest, and where the answers ramify the furthest.

Know thine enemy better than one knows thyself, Sun Tzu tells us in *The Art of War.* In the case of the Turing test, knowing our enemy actually *becomes* a way of knowing ourselves. So we will, indeed, have a look at how some of these bots are constructed, and at some of the basic principles and most important results in theoretical computer science, but always with our eye to the human side of the equation.

In a sense, this is a book about artificial intelligence, the story of its history and of my own personal involvement, in my own small way, in that history. But at the core, it's a book about living life.

We can think of computers, which take an increasingly central role in our lives, as nemeses: a force like *Terminator*'s Skynet, or *The Matrix*'s Matrix, bent on our destruction, just as we should be bent on theirs. But I prefer, for a number of reasons, the notion of *rivals*—who only ostensibly want to win, and who know that competition's main purpose is to raise the level of the game. All rivals are symbiotes. They need each other. They keep each other honest. They make each other better. The story of the progression of technology doesn't have to be a dehumanizing or dispiriting one. Quite, as you will see, the contrary.

In the months before the test, I did everything I could to prepare, researching and talking with experts in various areas that related back to the central questions of (a) how I could give the "most human" performance possible in Brighton, and (b) what, in fact, it means to be human. I interviewed linguists, information theorists, psychologists, lawyers, and philosophers, among others; these conversations provided both practical advice for the competition and opportunities to look at how the Turing test (with its concomitant questions of humanhood) affects and is affected by such far-flung fields as work, school, chess, dating, video games, psychiatry, and the law.

The final test, for me, was to give the most uniquely human performance I could in Brighton, to attempt a successful defense against the machines passing the test, and to take a run at bringing home the coveted, if bizarre, Most Human Human prize—but the ultimate question, of course, became what it *means* to be human: what the Turing test can teach us about ourselves.

2. Authenticating

Authentication: Form & Content

National Public Radio's *Morning Edition* recently reported the story of a man named Steve Royster. Growing up, Royster assumed he had an incredibly unusual and distinctive voice. As he explains, "Everyone always knew when I was calling just by the sound of my voice, while I had no earthly *idea* who was on the phone when *they* called." It would take him until his late twenties before he fully grasped—to his amazement—that other people could discern most *everyone's* identity by voice. How on earth could they do that? As it turns out, there *is* something unusual about Royster, but not about his voice: about his brain. Royster has a rare condition known as "phonagnosia," or "voice blindness." Even when Royster's own mother calls him, he simply goes politely along with the flow of the conversation, unaware that "this strange woman who has called me is, in fact, the one that gave birth to me." As reporter Alix Spiegel puts it, "Phonagnosics can tell from the sound of your voice if you're male or female, old or young, sarcastic, upset, happy. They just have no blooming idea who you are."

This all puts Royster, of course, in an awfully strange position.

It happens to be the same position everyone is in on the Internet.

On September 16, 2008, a twenty-year-old college student named David Kernell attempted to log in to vice-presidential candidate Sarah Palin's personal Yahoo! email account. He didn't have a clue what her password might be. Guessing seemed futile; instead, it occurred to him to try to *change* it—and so he clicked on the "I forgot my password" option available to assist absentminded users. Before Yahoo! will let a user change an account password, it asks the user to answer several "authentication" questions—things like date of birth and zip code—in order to "Verify Your Identity." Kernell found the information on Wikipedia, he said, in approximately "15 seconds." Stunned, Kernell "changed the password to 'popcorn' and took a cold shower." Now he faces up to twenty years in prison.

In the world of machines, we authenticate on *content*: password, PIN, last four digits of your Social Security number, your mother's maiden name. But in the human world, we authenticate on *form*: face, vocal timbre, handwriting, signature.

And, crucially, verbal style.

One of my friends emailed me recently: "I'm trying to rent a place in another city by email, and I don't want the fellow I've been communicating with to think I'm scamming him (or, am a flake), so, I've been hyperaware of sounding 'human' and 'real' and basically 'nonanonymous' in my emails. A weird thing. Do you know what I mean?" I do; it's *that* email's idiosyncrasies of style—the anachronistic "fellow," the compound, unhyphenated "hyperaware" and "nonanonymous"—that prove it's really *him*.

This kind of thing—behavior that seems "so you"—might always have been, say, charming or winning (at least to those who like you). Now it's something else too, our words increasingly dissociated from us in the era of the Internet: part of online *security*.[1]

---

1. When something online makes me think of a friend I haven't talked to in a while, and I want to send them a link, I make sure to add some kind of personal flourish, some little verbal fillip to the message beyond just the minimal "hey, saw this and thought of you / [*link*] / hope all's well," or else my message risks a spam-bin fate.

E.g., when I received the other week a short, generically phrased Twitter message from one of the poetry editors of *Fence* magazine saying, "hi, i'm 24/female/horny . . . i have to get off here but message me on my windows

Antarctic penguins detect the precise call of their chicks among the 150,000 families at the nesting site. "Bless Babel," fiction writer Donald Barthelme says. It's true: ironing out our idiosyncrasies in verbal style would not only be bad for literature; it would be bad for *safety*. Here as elsewhere, maybe that slight machine-exerted pressure to actively assert our humanity with each other ends up being a good thing.

Intimacy: Form & Content

One of my old college friends, Emily, came into town recently, and stopped downtown on her way from the airport to have lunch with a mutual friend of ours and his co-worker—who happened also to be my girlfriend, Sarah. When Emily and I met up later that day for dinner, I remarked on how funny it was that she'd already met Sarah before I'd had any chance to introduce them. I remember saying something to the effect of, "It's cool that you guys got to know each other a little bit." "Well, I wouldn't say that I got to *know* her, per se," Emily replied. "More like, 'saw what she's like' or something like that. 'Saw her in action.'"

And that's when the distinction hit me—

Having a *sense* of a person—their disposition, character, "way of being in the world"—and knowing *about* them—where they grew up, how many siblings they have, what they majored in, where they work—are two rather different things. Just like security, so does intimacy have both form and content.

"Speed dating" is a kind of fast-paced, highly structured round-robin-style social mixing event that emerged in Beverly Hills in the late 1990s. Each participant has a series of seven-minute conversations, and at the end they mark down on a card which people

---

live messenger name: [*link*]," my instinct wasn't to figure out how to politely respond that I was flattered but thought it best to keep our relationship professional; it was to hit the "Report Spam" button.

they'd be interested in meeting again; if there are any mutual matches, the organizers get in touch with the relevant contact information. Though it's entered into popular parlance, "SpeedDating" ("or any confusingly similar term") is technically a registered trademark, held by, of all groups, the Jewish organization Aish HaTorah: its inventor, Yaacov Deyo, is a rabbi.

One of my earliest thoughts about the Turing test was that it's a kind of speed date: you have five minutes to show another person who you are, to come across as a real, living, breathing, unique and distinct, nonanonymous human being. It's a tall order. And the stakes in both cases are pretty high.

A friend of mine recently went to a speed-dating event in New York City. "Well, it was the oddest thing," he said. "I kept wanting just to, like, banter, you know? To see if there was any chemistry. But all the women just kind of stuck to this script—where are you from, what do you do—like they were getting your stats, sizing you up. But I don't care about any of that stuff. So after a while I just started giving fake answers, just making stuff up, like. Just to keep it interesting, you know?"

The strangeness he experienced, and the kinds of "bullet points" that speed dating can frequently devolve into, are so well-known as to have been lampooned by *Sex and the City*:

> "Hi, I'm Miranda Hobbes."
> "Dwight Owens; private wealth group at Morgan Stanley; investment management for high-net-worth individuals and a couple pension plans; like my job; been there five years; divorced; no kids; not religious; I live in New Jersey; speak French and Portuguese; Wharton business school; any of this appealing to you?"

The delivery certainly isn't.

People with elaborate checklists of qualities their ideal mate must have frequently put entirely the wrong types of things. This height.

This salary. This profession. I've seen many a friend wind up, seemingly unsuspecting, with a jerk who nevertheless perfectly matched their description.

Fed up with the "Dwight Owens"–style, salvo-of-bullet-points approach that kept recurring in early speed-dating events, Yaacov Deyo decided on a simple, blunt solution: to make talking about your job *forbidden*. People fell back on talking about where they lived or where they were from. So he made that forbidden too. He seems charmed and maybe even a little smug enacting the ensuing panic, then breakthrough: "Omigosh, like, what *do* I talk about?" He laughs. "I can't talk about what I do for a living, or where I live, and . . . wow! All of a sudden I have to describe *me*." Or: all of a sudden I have to *be* me, to *act like* myself instead of *describing* myself.

Form and Content in the Turing Test

The first Loebner Prize competition was held on November 8, 1991, at the Boston Computer Museum. In its first few years, the Loebner Prize gave each program and human confederate a "topic" as a means of limiting the conversation. In some ways, that first contest was a battle between style and content. One of the confederates was Shakespeare expert Cynthia Clay, who was, famously, deemed a computer by three different judges. The consensus seemed to be "No one knows that much about Shakespeare." (For this reason, Clay took her misclassifications as a compliment.)

The program behind the topic of "whimsical conversation"—PC Therapist III by Joseph Weintraub—won the Loebner Prize that year, earning the very first Most Human Computer award. As the program practiced it, "whimsical conversation" was not a *topic* at all, but a *manner*—a crucial distinction. Weintraub was able to play off of a looseness in the contest rules, the natural behavior patterns of chatbots, and people's stereotypes about computers and humans all at once. Strategically, it was brilliant. By dutifully and scrupulously providing information in response to the questions asked, Clay demonstrated her knowledge and understanding—but sometimes in a factual, encyclopedic way commonly associated with computer systems.

> JUDGE: What is your opinion on Shakespeare's plays?
> REMOTE: That's pretty general; would you be more specific?
> Otherwise, I'll just say I like them.

The interlocutor at the "whimsical conversation" terminal stood in stark contrast:

> JUDGE: It has been a long day! Now I think I'm ready for some
> fun. The weather isn't very pleasant today. Do you have
> any jokes that could lift the clouds?
> REMOTE: Everybody talks about the weather but nobody
> seems to do much about it. Why do you need to tell
> me now you think you're ready for some fun?
> Groannnnnn.
> . . .
> JUDGE: What is the definition of whimsical conversation?
> REMOTE: Didn't you say well anyway I would imagine whimsical
> conversation is like a series of knight's moves in chess-
> skrt[2] of zig-zag rather than following any nparticular
> line.? Why do you ask? Was that a clever comment, or what?
> Are such questions much on your mind?
> JUDGE: I like the image of knights moving hap-hazardly across
> the chess board, does that mean there is no thought to
> whimsical conversation?
> REMOTE: I could swear you just said how do you define

---

2. *Sic.* Weintraub's program, like many that followed it, faked typos.

whimsical?? How do yku define whimsical? Were you always
so sick sick sick? Why do you need to tell me you like the
image of knights moving haphazardly across the chess board?

"I really thought [PC Therapist] was human because it . . . was
weird and funny, in a normal sort of way," said one of the judges,
a Harvard graduate student. Weintraub's program, shifting topics wildly and spouting non sequiturs and canned one-liners, came
off as zany, a jokester, a much more "human" personality type. At
least—before I learned how easy it was to mimic—I used to think so.

Suspicion; Roulette; Purée

Looking around on the Internet for information about the Most
Human Computer of 2005 and 2006, Rollo Carpenter's web-based
Cleverbot, I found something very curious. There seems to be a con-
siderable amount of speculation that the site may be, in effect, a hoax.

"I've been talking to it for a while now, and am beginning to wonder if Cleverbot is really even a real bot," one user writes. "Cleverbot
always says that itself is the human, and I am the bot. Could it be
that I'm actually talking to a real person, but we are both tricked to
believe that the other is Cleverbot?" Some of the responses seem
just a hair too uncanny. One user writes, in French, "Je habite dans
les Etats Unis," and finds his French not only comprehended but cri-
tiqued: "On dit J'habite AUX Etats-Unis," the bot corrects him.

Another trades off lines of Queen's "Bohemian Rhapsody":

USER: Scaramouche, Scaramouche.
CLEVERBOT: Will you do the fandango?
USER: Thunderbolt and lightning.
CLEVERBOT: Very very frightening me!
USER: Galileo.
CLEVERBOT: Galileo Galileo.

One user reports telling it a popular joke of the time—

> USER: You know why they call it an Xbox 360?
> CLEVERBOT: Why?
> USER: Because when you see it, you turn 360 degrees and walk away.

—only to have his joke deconstructed by the bot:

> CLEVERBOT: No because then you would be facing it and that would be unfunny.

Other users, attempting to fix the logic of the joke, meanwhile making it slightly more drily absurd—

> USER: You know why they call it an Xbox 360?
> CLEVERBOT: No. Why do they call it Xbox 360?
> USER: Because when you see it, you turn 360 degrees and moonwalk away.

—were met with a kind of one-upmanship of an even drier, more absurd punch line:

> CLEVERBOT: Nope, because when you see it you turn 3 degrees and walk slightly askew.

Surely there must be a human at the other side! The replies are too culturally attuned, too savvy. Yet at other moments, conversations change gears without a clutch, or a non sequitur comes seemingly from left field.

Cleverbot's programmer, Rollo Carpenter, is happy to explain his creation's programmatic workings, and insists on Cleverbot's home page that "visitors never talk to a human, however convincing it is." Curiously, this insistence seems to have little effect on many users, who have their own, rather different theory about what's going on.

The Internet of the early 1990s was a much more anonymous place than it is now. On local BBSs (bulletin board systems), in the chat rooms of "walled garden" Internet providers/communities like Prodigy and AOL, and over universal chat protocols like IRC (Internet Relay Chat), strangers bumped into each other all the time. The massive social networks (e.g., Facebook) of the late '00s and early '10s have begun to make the Internet a different place. It's around this time that websites like Chatroulette and Omegle, designed to bring some of that anonymity, randomness, and serendipity back, took off. You choose to use either video or text and are then paired up with another user completely at random and begin a conversation.[3] At any time, either of you can terminate it, in which case you're both re-paired with new strangers and begin instantly again at "Hello." There's an anxiety all users of such sites feel about the prospect of the other person cutting off the dialogue and bumping both of you into new conversations, which has been dubbed "getting nexted."

Now, imagine if, instead, the computer system was *automatically* cutting off conversations and re-pairing users with each other, and that it was *not telling them* it was doing this. Users A and B are arguing about baseball, and users C and D are talking about art. All of a sudden A is re-paired with C, and B re-paired with D. After talk-

---

3. Such anonymity brings hazard, though, at least as much as serendipity. I read someone's account of trying out Chatroulette for the first time: twelve of the first twenty video chats he attempted were with men masturbating in front of the camera. For this reason, and because it was more like the Turing test, I stuck to text. Still, my first two interlocutors on Omegle were guys trolling, stiltedly, for cybersex. But the third was a high school student from the suburbs of Chicago: we talked about *Cloud Gate,* the Art Institute, the pros and cons of growing up and moving out. Here was a real person. "You're normal!!" she wrote, with double exclamation marks; my thought exactly.

ing about the Louvre, C receives the off-topic "So are you for the Mets or the Yankees?" and B, after analyzing the most recent World Series, is asked if he's ever seen the Sistine Chapel. Well, this is the conspiracy theory on Cleverbot (and some of its cousin bots, like Robert Medeksza's Ultra Hal): Omegle minus control over when to switch conversations. Imagine that the computer is simply switching you over, at random and without notice, to new people, and doing the same to them. What you'd end up with might look a lot like the Cleverbot transcripts.

The conspiracy theory isn't right, but it's not far off either.

"Cleverbot borrows the intelligence of its users," Carpenter explains to me in Brighton. "A conversational Wikipedia," he calls it in a television interview with the Science Channel. It works like this: Cleverbot begins a conversation by saying, for instance, "Hello." A user might respond in any number of ways, from "Hello" to "Howdy!" to "Are you a computer?" and so on. Whatever the user says goes into an enormous database of utterances, tagged as a genuine human response to "Hello." When, in a subsequent conversation, a user ever says to Cleverbot, "Hello," Cleverbot will have "Howdy!" (or whatever the first person said) ready on hand. As the same types of things tend to come up over and over—in what statisticians call a "Zipf distribution," to be precise—and as thousands of users are logged in to Cleverbot at any given time, chatting with it around the clock, over the span of many years now, Cleverbot's database contains appropriate replies to even seemingly obscure remarks. (E.g., "Scaramouche, Scaramouche.")

What you get, the cobbling together of hundreds of thousands of prior conversations, is a kind of conversational purée. Made of human parts, but less than a human sum. Users *are*, in effect, chatting with a kind of purée of real people—the *ghosts* of real people, at any rate: the echoes of conversations past.

This is part of why Cleverbot seems so impressive on basic factual questions ("What's the capital of France?" "Paris is the capital of France") and pop culture (trivia, jokes, and song lyric sing-alongs)—the things to which there is a *right* answer independent of the speaker. No number of cooks can spoil the broth. But ask it about the city it lives in, and you get a pastiche of thousands of people talking about thousands of places. You find it out not so much by realizing that you aren't talking with a *human* as by realizing that you aren't talking with *a* human.

One Self, Any Self

My mind goes back to the organizers' advice to "just be myself," to how much philosophers have agonized over this idea. While other existentialists—for instance, Jean-Paul Sartre—emphasized authenticity and originality and freedom from outside influence, nineteenth-century German philosopher Friedrich Nietzsche held the startling opinion that the most important part of "being oneself" was—in Brown University philosopher Bernard Reginster's words—"being *one* self, *any* self."

Nietzsche spoke of this as "giving style to one's character," comparing people to works of art, which we often judge according to their "concinnity," the way their parts fit together to make a whole: "In the end, when the work is finished, it becomes evident how the constraint of a single taste governed and formed everything large and small."

Computer culture critics like Jaron Lanier are skeptical, for instance, of decentralized projects like Wikipedia, arguing:

> *The Sims,* . . . the iPhone, the Pixar movies, and all the other beloved successes of digital culture . . . are personal expressions. True, they often involve large groups of collaborators, but there is always a central personal vision—a Will Wright, a Steve Jobs, or a Brad Bird conceiving the vision and directing a team of people earning salaries.

It is this same "central personal vision" that is crucial for Nietzsche, who goes so far as to say, "Whether this taste was good or bad is less important than one might suppose, if only it was a single taste!"

It is precisely the "central personal vision" of Lanier and "single taste" of Nietzsche that is lacking in most chatbots. For instance, I had the following conversation with "Joan," the Cleverbot-offshoot program that won the Loebner Prize in 2006. Though each of her answers, taken separately, is perfectly sensible and human, their sum produces nothing but a hilarious cacophony in the way of identity:

```
USER: Do you have a boyfriend?
JOAN: Not at the moment, but I hope to find one soon.
```

I ventured:

```
USER: I would like to be your boyfriend.
JOAN: That would be difficult, since I'm happily married.
```

Curious. I asked again, for clarification:

```
USER: Do you have a boyfriend?
JOAN: No, I'm still looking for love.
USER: Do you have a husband?
JOAN: I don't have a husband, because it would be illegal for
    me to have one, since I am male.
```

Male?

```
USER: What gender are you?
JOAN: Female.
```

And once again, just for good measure:

```
USER: Do you have a boyfriend?
JOAN: Not at the moment, but I live in hope.
USER: Can I be your boyfriend?
JOAN: No, I'm straight.
```

This kind of unity or coherence of identity is something that most humans, of course—being the products of a single and continuous life history—have. But given the extreme brevity of a five-minute conversation, displaying that kind of congruence was something I tried to be aware of. For instance, when a judge said hello to my fellow confederate Dave, Dave replied with the nicely colorful and cheerful "G'day mate."

The drawback of this choice becomes immediately clear, however, as the judge's next question was "Have you come far to be here?" The judge, I imagine, was expecting some reference to Australia, the land that "G'day mate" evokes; instead, Dave answered, "From the southwest US." To the judge's mild surprise, I imagine, he discovers that Dave is not Australian at all, as his salutation would suggest, but rather an American from Westchester, New York, living in Albuquerque. It's not game over—it doesn't take Dave too long to win over the judge's confidence (and his vote)—but those signs of disjointed identity are early warning flags and, in that sense, falters.

In similar fashion, when a judge I was talking to spelled "color" in the British style ("colour"), and then several messages later referenced "Ny," which I took to mean "New York" (actually it turned out to be a typo for "My"), I asked where he was from. "Canadian spelling, not Biritish [sic]," he explained; my hope was that showing attunement, and over multiple utterances, to these questions of cohesiveness of identity would help my case. Presumably, a bot that can't keep track of the coherence of its *own* identity wouldn't be able to keep track of the judge's either.

"When making a bot, you don't write a program, you write a novel," explain programmers Eugene Demchenko and Vladimir Veselov, whose program "Eugene Goostman" was the runner-up at the 2008 competition, as well as in 2005 and 2001. They stress the importance of having a single programmer write the machine's responses: "Elect who will be responsible for the bot personality. The knowledge-base writing process can be compared to writing a book. Suppose every developer describes an episode without having any information on the others. Can you imagine what will be produced!"

In fact, it's quite easy to imagine what will be produced: "Eugene Goostman"'s competitors. This is a central trade-off in the world of bot programming, between coherence of the program's personality or style and the range of its responses. By "crowdsourcing" the task of writing a program's responses to the users themselves, the program acquires an explosive growth in its behaviors, but these behaviors stop being internally consistent.

Death of the Author; End of the Best Friend

Do you need someone? Or do you need me?

— SAY ANYTHING . . .

Speaking of "writing a book": this notion of style versus content, and of singularity and uniqueness of vision, is at the heart of recent debates about machine translation, especially of literature.

Wolfram Alpha researcher and chatbot author Robert Lockhart describes the chatbot community as being split between two competing approaches, what he calls "pure semantics" and "pure empiricism." Roughly speaking, the semantic camp tries to program linguistic *understanding*, with the hope that the desired behavior will follow, and the empirical camp tries to directly program linguistic *behavior*, with the hope that "understanding" will either happen along the way or prove to be an unnecessary middleman. This divide also plays out in the history of computer translation. For many decades, machine translation projects attempted to understand language in a rule-based way, breaking down a sentence's structure and getting down to the underlying, universal meaning, before re-encoding that meaning according to another language's rules. In the 1990s, a statistical approach to machine translation—the approach that Google uses—came into its own, which left the question of meaning entirely out of it.

Cleverbot, for instance, can know that "Scaramouche, Scaramouche" is best answered by "Will you do the fandango?" without needing any links to Queen or "Bohemian Rhapsody" in between, let alone needing to know that Scaramouche is a stock character in seventeenth-century Italian farce theater and that the fandango is an Andalusian folk dance. It's simply observed people saying one, then the other. Using huge bodies of text ("corpora") from certified United Nations translators, Google Translate and its statistical cousins regurgitate previous human translations the way Cleverbot and its cousins regurgitate previous human speech. Both Google Translate and Cleverbot show weaknesses for (1) unusual and/or nonliteral phrasing, and (2) long-term consistency in point of view and style. On both of those counts, even as machine translation increasingly penetrates the world of business, literary novels remain mostly untranslatable by machine.

What this also suggests, intriguingly, is that the task of translating (or writing) literary novels cannot be broken into parts and done by a succession of different *humans* either—not by wikis, nor crowdsourcing, nor ghostwriters. Stability of point of view and consistency of style are too important. What's truly strange, then, is the fact that we *do* seem to make a lot of art this way.

To be human is to be *a* human, a specific person with a life history and idiosyncrasy and point of view; artificial intelligence suggests that the line between intelligent machines and people blurs most when a purée is made of that identity. It is profoundly odd, then—especially so in a country with a reputation for "individualism"—to contemplate how often we do just that.

The British television series *The Office* consists of fourteen episodes, all written and directed by the two series creators, Ricky Gervais and Stephen Merchant. The show was so successful that it was spun off into an American version: 130 episodes and counting, each written by a different person from the last and each directed by a different person from the last. The only thing stable from week to week seems to be the cast. The arts in America are strange that way: we seem to care what our vision falls upon, but not whose vision it is.

I remember being enchanted as a kid with the early Hardy Boys books by Franklin W. Dixon, but after a certain point in the series, the magic seemed to disappear. It wasn't until more than fifteen years later I discovered that Franklin W. Dixon never existed. The first sixteen books were written by a man named Leslie McFarlane. The next twenty were written by eleven different people. What I'd chalked up to the loss of something intangible in those later books was in fact the loss of something very tangible indeed: the author.

Aesthetic experiences like these for me are like an unending series of blind dates where you never follow up, conversations with a stranger on the bus (or the Internet) where you never catch the other person's name. There's nothing *wrong* with them—they're pleasant, sometimes memorable, even illuminating—and all relationships start somewhere. But to live a whole *life* like that?

The *New York Times* reported in June 2010—in an article titled "The End of the Best Friend"—on the practice of deliberate intervention, on the part of well-meaning adults, to disrupt close nuclei of friends from forming in schools and summer camps.[4] One sleepaway camp in New York State, they wrote, has hired "friendship coaches" whose job is to notice whether "two children seem to be too focused on each other, [and] . . . put them on different sports teams [or] seat them at different ends of the dining table." Affirms one school counselor in St. Louis, "I think it is kids' preference to pair up and have that one best friend. As adults—teachers and counselors—we try to encourage them not to do that." Chatroulette and Omegle users "next" each other when the conversation flags; these children are being nexted by force—when things are going too *well*.

---

4. Motives range from wanting the children not to put all of their emotional eggs in one basket, to wanting them to branch out and experience new perspectives, to reducing the occasionally harmful social exclusion that can accompany tight bonds.

Nexted in Customer Service

The same thing happens sometimes in customer service, where the disruption of intimacy seems almost tactical. Recently a merchant made a charge to my credit card in error, which I attempted to clear up, resulting in my entering a bureaucratic Rube Goldberg machine the likes of which I had never before experienced. My record for the longest single call was forty-two minutes and *eight transfers*.

The ultimate conclusion reached at the end of this particular call was "call back tomorrow."

Each call, each transfer, led me to a different service rep, each of whom was skeptical and testy about the validity of my refund request. If I managed to get a particular rep on my side, to earn their sympathy, to start to build a kind of relationship and come across as a distinct "nonanonymous" human being, it was only a few minutes before I'd be talking to someone else, anonymous again. Here's my name, here's my account number, here's my PIN, here's my Social, here's my mother's maiden name, here's my address, here's the reason for my call, yes, I've already tried that . . .

What a familiarity with the construction of Turing test bots had begun showing me was that we fail—again and again—to actually *be* human with other humans, so maddeningly much of the time. And it had begun showing me *how* we fail—and what to do about it.

Cobbled-together bits of human interaction do not a human relationship make. Not fifty one-night stands, not fifty speed dates, not fifty transfers through the bureaucratic pachinko. No more than sapling tied to sapling, oak though they may be, makes an oak. Fragmentary humanity isn't humanity.

The Same Person

If the difference between a conversational purée and a conversation is continuity, then the solution, in this case, is extraordinarily simple:

assign a rep to a case. A particular person sees it through from start to finish. The *same* person.

For a brief period a tiny plastic tab that held the SIM card in my phone had gotten loose, and so my phone only worked when I was pressing on this plastic tab with my finger. As a result, I could only make calls, not receive them. And if I took my finger off the tab mid-call, the call dropped.

The tab is little more valuable than the plastic equivalent of a soda can's pull tab, which it resembles in appearance, and is roughly as essential for the proper functioning of the device it's attached to. I was out of warranty; protocol was that I was out of luck and needed a new, multi-hundred-dollar phone. "But this tab weighs one gram and costs a penny to manufacture," I said. "I know," said the customer service rep.

There was *no* way, no way at all, I couldn't just purchase a tab from them?

"I don't think it will work," she said. "But let me talk to a manager."

Then *the same woman* got back on the line. "I'm sorry," she said. "But . . ." I said. And we kept talking. "Well, let me talk to a *senior* manager, hold on," she says.

As I'm holding, I feel my hand, which has now been pushing down steadily on the plastic tab for about fifteen minutes, begin to cramp. If my finger slips off the tab, if she hits the wrong button on her console, if there is some glitch in my phone provider's network, or hers—I am anonymous again. Anybody. A nobody. A number. This particular person and I will never reconnect.

I must call again, introduce myself again, explain my problem again, hear again that protocol is against me, plead my case again.

Service works by the gradual buildup of sympathy through failed attempted solutions. If person X has told you to try something and it doesn't work, person X feels slightly sorry for you. X is slightly *responsible* for the problem now, having used up some of your time. Person Y, however, is considerably less moved that you tried following her colleague X's advice to no avail—even if it is the same advice that she herself would have given you had she been party to that earlier conversation. That's beside the point. The point is that she wasn't the one who gave you that advice. So she is not responsible for your wasted time.

The *same* woman, as if miraculously, again returns. "I can make an exception for you," she says.

It occurs to me that an "exception" is what programmers call it when software breaks.

50 First Dates

Sometimes even a single, stable point of view, a unifying vision and style and taste, isn't enough. You also need a *memory*. In the 2004 comedy *50 First Dates,* Adam Sandler courts Drew Barrymore, but in the process discovers that due to an accident she can't form new long-term memories.

Philosophers interested in friendship, romance, and intimacy more generally have, in recent times, endeavored to distinguish between the *types* of people we like (or, the things we like *about* people) and the *specific* people we feel connections with in our lives. University of Toronto philosopher Jennifer Whiting has dubbed the former "impersonal friends." The difference between the numerous "impersonal friends" out there, who are more or less fungible, and the few individuals we care about *specifically,* who aren't fungible with anyone on the planet, lies, she says, in so-called "historical properties." Namely, your actual friends and your innumerable "impersonal friends" *are* fungible—but only at the moment the relationship begins. From there, the relationship puts down roots, builds up a shared history, shared understanding, shared experiences, sacrifices and compromises and triumphs . . .

Barrymore and Sandler really *are* good together—life-partner good—but she becomes "someone special" to him, whereas he is doomed to remain merely "her type." Fungible. And therefore—being no different from the *next* charming and stimulating and endearing guy who shows up at her restaurant—*vulnerable* to losing her.

His solution: give her a historical-properties crash course every morning, in the form of a video primer that recaps their love. He must fight his way out of fungibility every morning.

Statefulness

A look at the "home turf" of many chatbots shows a conscious effort on the part of the programmers to make Drew Barrymores of us: worse, actually, because it was her *long*-term memory that kept wiping clean. At 2008 Loebner Prize winner Elbot's website, the screen refreshes each time a new remark is entered, so the conversational history evaporates with each sentence; ditto at the page of 2007 winner Ultra Hal. At the Cleverbot site, the conversation fades to white above the box where text is entered, preserving only the last three exchanges on the screen, with the history beyond that gone: out of sight, and hopefully—it would seem—out of the user's mind as well. The elimination of the long-term influence of conversational history makes the bots' jobs easier—in terms of both the psychology and the mathematics.

In many cases, though, physically eliminating the conversation log is unnecessary. As three-time Loebner Prize winner ('00, '01, and '04), programmer Richard Wallace explains, "Experience with [Wallace's chatbot] A.L.I.C.E. indicates that most casual conversation is 'state-less,' that is, each reply depends only on the current query, without any knowledge of the history of the conversation required to formulate the reply."

Not all types of human conversations function in this way, but many do, and it behooves AI researchers to determine which types of conversations are "stateless"—that is, with each remark depending only on the last—and to attempt to create these very sorts of interactions. It's our job as confederates, as humans, to resist it.

One of the classic stateless conversation types, it turns out, is verbal abuse.

In 1989, twenty-one-year-old University College Dublin undergraduate Mark Humphrys connects a chatbot program he'd written called MGonz to his university's computer network and leaves the building for the day. A user (screen name "SOMEONE") from Drake University in Iowa tentatively sends the message "finger" to Humphrys's account—an early-Internet command that acts as a request for basic information about a user. To SOMEONE's surprise, a response comes back immediately: "cut this cryptic shit speak in full sentences." This begins an argument between SOMEONE and MGonz that will last almost an hour and a half.

(The best part is undoubtedly when SOMEONE says, a mere twenty minutes in, "you sound like a goddamn robot that repeats everything.")

Returning to the lab the next morning, Humphrys is stunned to find the logs, and feels a strange, ambivalent emotion. His program might have just passed the Turing test, he thinks—but the evidence is so profane that he's afraid to publish it.

Humphrys's twist on the age-old chatbot paradigm of the "non-directive" conversationalist who lets the user do all the talking was to model his program, rather than on an attentive listener, on an abusive jerk. When it lacks any clear cue for what to say, MGonz falls back not on therapy clichés like "How does that make you feel?" or "Tell me more about that" but on things like "you are obviously an asshole," "ok thats it im not talking to you any more," or "ah type something interesting or shut up." It's a stroke of genius, because, as becomes painfully clear from reading the MGonz transcripts, *argument is stateless.*

I've seen it happen between friends: "Once again, you've neglected to do what you've promised." "Oh, there you go right in with that tone of yours!" "Great, let's just dodge the issue and talk about my tone instead! You're so defensive!" "*You're* the one being defensive! This is just like the time you *x*!" "For the millionth time, I did not even remotely x! *You're* the one who . . ." And on and on. A close reading of this dialogue, with MGonz in mind, turns up something interesting, and very telling: each remark after the first is *only about the previous remark*. The friends' conversation has become stateless, unanchored from all context, a kind of "Markov chain" of riposte, meta-riposte, meta-meta-riposte. If we can be induced to sink to this level, of course the Turing test can be passed.

Once again, the scientific perspective on what types of human behavior are imitable shines incredible light on how we conduct our own, human lives. There's a sense in which verbal abuse is simply *less complex* than other forms of conversation. Seeing how much MGonz's arguments resemble our own might shame us into shape.

Retorts, no matter how sharp or stinging, play into chatbots' hands. In contrast, requests for elaboration, like "In what sense?" and "How so?" turn out to be crushingly difficult for many bots to handle: because elaboration is hard to do when one is working from a prepared script, because such questions rely *entirely* on context for their meaning, and because they extend the relevant conversational history, rather than resetting it.

In fact, since reading the papers on MGonz, and its transcripts, I find myself much more able to constructively manage heated conversations. Aware of their stateless, knee-jerk character, I recognize that the terse remark I want to blurt has far more to do with some kind of "reflex" to the very last sentence of the conversation than it does with either the actual issue at hand or the person I'm talking to. All of a sudden the absurdity and ridiculousness of this kind of escalation become *quantitatively* clear, and, contemptuously unwilling to act like a bot, I steer myself toward a more "stateful" response: better living through science.

3. The Migratory Soul

I'm Up Here

The Turing test attempts to discern whether computers are, to put it most simply, "like us" or "unlike us": humans have always been preoccupied with their place among the rest of creation. The development of the computer in the twentieth century may represent the first time that this place has changed.

The story of the Turing test, of the speculation and enthusiasm and unease over artificial intelligence in general, is, then, the story of our speculation and enthusiasm and unease over ourselves. What are our abilities? What are we good at? What makes us special? A look at the history of computing technology, then, is only half of the picture. The other half is the history of mankind's thoughts about itself. This story takes us back through the history of the soul, and it begins at perhaps the unlikeliest of places, that moment when the woman catches the guy glancing at her breasts and admonishes him: "Hey—I'm up *here*."

Of course we look each other in the eyes by default—the face is the most subtly expressive musculature in the body, for one, and knowing where the *other* person is looking is a big part of communication (if their gaze darts to the side inexplicably, we'll perk up and look there too). We look each other in the eyes and face because we care about what the other person is feeling and thinking and attending to, and so to ignore all this information in favor of a mere ogle is, of course, disrespectful.

In fact, humans are known to have the largest and most visible sclera—the "whites" of the eyes—of any species. This fact intrigues scientists, because it would seem actually to be a considerable hindrance: imagine, for example, the classic war movie scene where the soldier dresses in camouflage and smears his face with green and brown pigment—but can do nothing about his conspicuously white sclera, beaming bright against the jungle. There must be *some* reason humans developed it, despite its obvious costs. In fact, the advantage of visible sclera—so goes the "cooperative eye hypothesis"—is precisely that it enables humans to see clearly, and from a distance, which direction other humans are looking. Michael Tomasello at the Max Planck Institute for Evolutionary Anthropology showed in a 2007 study that chimpanzees, gorillas, and bonobos—our nearest cousins—follow the direction of each other's *heads,* whereas human infants follow the direction of each other's *eyes.* So the value of looking someone in the eye may in fact be something uniquely human.

But—this happens not to be the woman's argument in this particular case. Her argument is that *she's* at eye level.

As an informal experiment, I will sometimes ask people something like "Where are you? Point to the exact place." Most people point to their forehead, or temple, or in between their eyes. Part of this must be the dominance, in our society anyway, of the sense of vision—we tend to situate ourselves at our visual point of view—and part of it, of course, comes from our sense, as twenty-first-centuryites, that the *brain* is where all the action happens. The mind is "in" the brain. The soul, if anywhere, is there too; in fact, in the seventeenth century, Descartes went so far as to try to hunt down the *exact* "seat of the soul" in the body, reckoning it to be the pineal gland at the center of the brain. "The part of the body in which the soul directly exercises its functions[1] is not the heart at all, or the whole of the brain," he writes. "It is rather the innermost part of the brain, which is a certain very small gland."[2]

Not the heart at all—

Descartes's project of trying to pinpoint the exact location of the soul and the self was one he shared with any number of thinkers and civilizations before him, but not much was thought of the brain for most of human history. The ancient Egyptian mummification process involved, for instance, preserving all of a person's organs *except* the brain—thought[3] to be useless—which they scrambled with hooks into a custard and scooped out through the nose. All the other major organs—stomach, intestines, lungs, liver—were put into sealed jars, and the heart alone was left in the body, because it was considered, as Carl Zimmer puts it in *Soul Made Flesh,* "the center of the person's being and intelligence."

In fact, *most* cultures have placed the self in the thoracic region somewhere, in one of the organs of the chest. This historical notion of heart-based thought and feeling leaves its fossil record in the idioms and figurative language of English: "that shows a lot of heart," we

---

1. When I first read about this as an undergraduate, it seemed ridiculous, the notion that a nonphysical, nonspatial entity like the soul would somehow *deign* to physicality/locality in order to "attach" itself to the physical, spatial brain at any specific point—it just seemed ridiculous to try to *locate* something *non-localized.* But later that semester, jamming an external wireless card into my old laptop and hopping online, I realized that the idea of accessing something vague, indefinite, all surrounding, and un-locatable—my first reaction to my father explaining how he could "go to the World Wide Web" was to say, "Where's that?"—through a specific physical component or "access point" was maybe not so prima facie laughable after all.

2. Depending on your scientific and religious perspectives, the soul/body interface might have to be a special place where normal, deterministic cause-and-effect physics breaks down. This is metaphysically awkward, and so it makes sense that Descartes wants to shrink that physics-violation zone down as much as possible.

3. !

say, or "it breaks my heart," or "in my heart of hearts." In a number of other languages—e.g., Persian, Urdu, Hindi, Zulu—this role is played by the liver: "that shows a lot of liver," their idioms read. And the Akkadian terms *karšu* (heart), *kabattu* (liver), and *libbu* (stomach) all signified, in various different ancient texts, the center of a person's (or a deity's) thinking, deliberation, and consciousness.

I imagine an ancient Egyptian woman, say, who catches a man looking tenderly into her eyes, up at the far extreme of her body near her useless, good-for-nothing brains, and chastises him, hand at her chest. Hey. I'm down *here*.

A Brief History of the Soul

The meaning and usage of the word "soul" in ancient Greece ($\psi\upsilon\chi\acute{\eta}$—written as "psyche"[4]) changes dramatically from century to century, and from philosopher to philosopher. It's fairly difficult to sort it all out. Of course people don't speak in twenty-first-century America the way they did in nineteenth-century America, but scholars of the next millennium will have a hard time becoming as sensitive to those differences as we are. Even differences of *four* hundred years are sometimes tricky to keep in mind: when Shakespeare writes of his beloved that "black wires grow on her head," it's easy to forget that electricity was still several centuries away. He's not likening his lover's hair to the shelves of RadioShack. And smaller and more nuanced distinctions are gnarlier by far. "Hah, that's so '80s," we sometimes said to our friends' jokes, as early as the '90s . . . Can you imagine looking at a text from 460 B.C. and realizing that the author is talking *ironically* like someone from 470 B.C.?

Back to "soul": the full story runs long, but a number of fascinat-

---

4. The word "psyche" has, itself, entered English as a related, but not synonymous, term to "soul"—one of the many quirks of history that make philology and etymology so convoluted and frustrating and interesting.

ing points are raised at various moments in history. In Plato's *Phaedo* (360 B.C.), Socrates, facing his impending execution, argues that the soul is (in scholar Hendrik Lorenz's words) "less subject to dissolution and destruction than the body, rather than, as the popular view has it, more so." *More* so! This fascinated me to read. Socrates was arguing that the soul somehow *transcended* matter, whereas his countrymen, it would seem, tended to believe that the soul was made of a supremely gossamer, delicate, fine form of matter[5]—this was Heraclitus's view[6]—and was therefore *more* vulnerable than the meatier, hardier tissues of the body. Though at first the notion of a fragile, material soul seems ludicrously out of line with everything we traditionally imagine about the soul, it makes more sense of, if offers less consolation for, things like head injury and Alzheimer's. Likewise, part of the debate over abortion involves the question of when, exactly, a person *becomes* a person. The human body, Greeks of the fourth century B.C. believed, can both pre- and postdate the soul.

Along with questions of the composition and durability of the soul came questions of who and what had them. It's not just the psychologists who have been invested in The Sentence: philosophers, too, seem oddly riveted on staking out just exactly what makes *Homo sapiens* different and unique. Though Homer only used the word "psyche" in the context of humans, many of the thinkers and writers that followed him began to apply it considerably more liberally. Empedocles, Anaxagoras, and Democritus referred to plants and animals with the same word; Empedocles believed he was a bush in a previous life; Thales of Miletus suspected that magnets, because they had the power to move other objects, might have souls.

Oddly, the word appears to have been used both more broadly and

---

5. Fine indeed. "A piece of your brain the size of a grain of sand would contain one hundred thousand neurons, two million axons, and one billion synapses, all 'talking to' each other."
6. Philolaus's different but related view was that the soul is a kind of "attunement" of the body.

more narrowly than it tends to be used in our culture today. It's used to describe a general kind of "life force" that animates everything from humans to grasses, but it's also construed specifically quite intellectually. In the *Phaedo*, the earlier of Plato's two major works on the soul, Socrates ascribes beliefs, pleasures, desires, and fears to the *body*, while the soul is in charge of regulating these and of "grasping truth."

In Plato's later work, *The Republic*, he describes the soul as having three distinct parts—"appetite," "spirit," and "reason"—with those first two "lower" parts taking those duties (hunger, fear, and the like) from the body.

Like Plato, Aristotle didn't believe that people had a soul—he believed we had three. His three were somewhat different from Plato's, but they match up fairly well. For Aristotle, all plants and animals have a "nutritive" soul, which arises from biological nourishment and growth, and all animals additionally have an "appetitive" soul, which arises from movement and action. But humans alone had a third, "rational" soul.

I say "arises from" as opposed to "governs" or something along those lines; Aristotle was quite interesting in this regard. For him the soul was the *effect* of behavior, not the *cause*. Questions like this continue to haunt the Turing test, which ascribes intelligence purely on the basis of behavior.

After Plato and Aristotle came a school of Greek philosophy called Stoicism. Stoics placed the mind at the heart, and appear to have taken a dramatic step of severing the notion of the "soul" from the notion of life in general: for them, unlike for Plato and Aristotle, plants did *not* have souls. Thus, as Stoicism ascended to popularity in Greece, the soul became no longer responsible for life function in general, but specifically for its mental and psychological aspects.[7]

---

7. The Stoics had another interesting theory, which foreshadows nicely some of the computer science developments of the 1990s. Plato's theory of the tripartite soul could make sense of situations where you feel ambivalent or "of

No Dogs Go to Heaven

Stoicism appears to have been among the tributary philosophies that fed into Christianity, and which also led to the seminal philosophical theories of mind of René Descartes. For the monotheistic Descartes, presumably the (Platonic) notion of multiple souls crowding around was a bit unsavory (although who could deny the Christian appeal of the three-in-one-ness?), and so he looked to draw that us-and-them line using just a single soul, *the* soul. He went remarkably further than Aristotle, saying, in effect, that all animals besides humans don't have *any* kind of soul at *all*.

Now, any kid who grows up going to Sunday school knows that this is a touchy point of Christian theology. All kids ask uncomfortable questions once their pets start to die, and tend to get relatively awkward or ad hoc answers. It comes up all over the place in mainstream culture too, from the deliberately provocative title of *All Dogs Go to Heaven* to the wonderful moment in *Chocolat* when the new priest, tongue-tied and flummoxed by a parishioner's asking whether it was sinful for his (soulless) dog to enter a sweet shop during Lent, sum-

---

two minds" about something—he could describe it as a clash between two different parts of the soul. But the Stoics only had one soul with one set of functions, and they took pains to describe it as "indivisible." How, then, to explain ambivalence? In Plutarch's words, it is "a turning of the single reason in both directions, which we do not notice owing to the sharpness and speed of the change." In the '90s, I recall seeing an ad on television for Windows 95, where four different animations were onscreen, playing one at a time as a mouse pointer clicked from each to the next. This represented old operating systems. All of a sudden all four animations began running simultaneously: this represented Windows 95, with *multitasking*. Until around 2007 and onward, when multiprocessor machines became increasingly standard, multitasking was simply—Stoic-style—switching back and forth between processes, just as with the old operating systems the ad disparages, except doing so automatically, and really fast.

marily prescribes some Hail Marys and Our Fathers and slams the confessional window. End of discussion.

Where some of the Greeks had imagined animals and even plants as "ensouled"—Empedocles thinking he'd lived as a bush in a past life—Descartes, in contrast, was firm and unapologetic. Even Aristotle's idea of multiple souls, or Plato's of partial souls, didn't satisfy him. Our proprietary, uniquely human soul was the only one. No dogs go to heaven.

The End to End All Ends: Eudaimonia

Where is all this soul talk going, though? To describe our animating force is to describe our nature, and our place in the world, which is to describe how we ought to live.

Aristotle, in the fourth century B.C., tackled the issue in *The Nicomachean Ethics*. The main argument of *The Nicomachean Ethics*, one of his most famous works, goes a little something like this. In life there are means and ends: we do x so that y. But most "ends" are just, themselves, means to other ends. We gas up our car to go to the store, go to the store to buy printer paper, buy printer paper to send out our résumé, send out our résumé to get a job, get a job to make money, make money to buy food, buy food to stay alive, stay alive to . . . well, what, exactly, is the goal of *living*?

There's one end, only one, Aristotle says, which doesn't give way to some other end behind it. The name for this end, εὐδαιμονία in Greek—we write it "eudaimonia"—has various translations: "happiness" is the most common, and "success" and "flourishing" are others. Etymologically, it means something along the lines of "well-being of spirit." I like "flourishing" best as a translation—it doesn't allow for the superficially hedonistic or passive pleasures that can sometimes sneak in under the umbrella of "happiness" (eating Fritos often makes me "happy," but it's not clear that I "flourish" by doing so), nor the superficially competitive and potentially cutthroat aspects of "success" (I might "succeed" by beating my middle school classmate at paper football, or by getting away with massive investor fraud, or by killing a rival in a duel, but again, none of these seems to have much to do with "flourishing"). Like the botanical metaphor underneath it, "flourishing" suggests transience, ephemerality, a kind of process-over-product emphasis, as well as the sense—which is crucial in Aristotle—of doing what one is meant to do, fulfilling one's promise and potential.

Another critical strike against "happiness"—and a reason that it's slightly closer to "success"—is that the Greeks don't appear to care about what you actually *feel*. Eudaimonia is eudaimonia, whether you recognize and experience it or not. You can think you have it and be wrong; you can think you *don't* have it and be wrong.[8]

Crucial to eudaimonia is ἀρετή—"arete"—translated as "excellence" and "fulfillment of purpose." Arete applies equally to the organic and the inorganic: a blossoming tree in the spring has arete, and a sharp kitchen knife chopping a carrot has it.

To borrow from a radically different philosopher—Nietzsche—"There is nothing better than what is good! and that is: to have a certain kind of capacity and to use it." In a gentler, slightly more botanical sense, this is Aristotle's point too. And so the task he sets out for himself is to figure out the capacity of humans. Flowers are meant to bloom; knives are meant to cut; what are we meant to do?

Aristotle's Sentence; Aristotle's Sentence Fails

Aristotle took what I think is a pretty reasonable approach and decided to address the question of humans' purpose by looking at

---

8. This is an interesting nuance, because of how crucial the subjective/objective distinction has been to modern philosophy of mind. In fact, subjective experience seems to be the linchpin, the critical defensive piece, in a number of arguments against things like machine intelligence. The Greeks didn't seem too concerned with it.

what capacities they had that animals lacked. Plants could derive nourishment and thrive physically; animals seemed to have wills and desires, and could move and run and hunt and create basic social structures; but only humans, it seemed, could *reason*.

Thus, says Aristotle, the human arete lies in contemplation—"perfect happiness is a kind of contemplative activity," he says, adding for good measure that "the activity of the gods . . . must be a form of contemplation." We can only imagine how unbelievably convenient a conclusion this is for a *professional philosopher* to draw—and we may rightly suspect a conflict of interest. Then again, it's hard to say whether his conclusions derived from his lifestyle or his lifestyle derived from his conclusions, and so we shouldn't be so quick to judge. Plus, who among us wouldn't have some self-interest in describing their notion of "the most human human"? Still, despite the grain of salt that "thinkers' praise of thinking" should have been taken with, the emphasis they placed on reason seemed to stick.

The Cogito

The emphasis on reason has its backers in Greek thought, not just with Aristotle. The Stoics, as we saw, also shrank the soul's domain to that of reason. But Aristotle's view on reason is tempered by his belief that sensory impressions are the currency, or language, of thought. (The Epicureans, the rivals of the Stoics, believed sensory experience—what contemporary philosophers call *qualia*—rather than intellectual thought, to be the distinguishing feature of beings with souls.) But Plato seemed to want as little to do with the actual, raw experience of the world as possible, preferring the relative perfection and clarity of abstraction, and, before him, Socrates spoke of how a mind that focused too much on sense experience was "drunk," "distracted," and "blinded."[9]

---

9. In Hendrik Lorenz's words: "When the soul makes use of the senses and attends to perceptibles, 'it strays and is confused and dizzy, as if it were

Descartes, in the seventeenth century, picks up these threads and leverages the mistrust of the senses toward a kind of radical skepticism: How do I know my hands are really in front of me? How do I know the world actually exists? How do I know that *I* exist?

His answer becomes the most famous sentence in all of philosophy. *Cogito ergo sum.* I think, therefore I am.

I *think,* therefore I am—not "I register the world" (as Epicurus might have put it), or "I experience," or "I feel," or "I desire," or "I recognize," or "I sense." No. I *think.* The capacity furthest *away* from lived reality is that which assures us of lived reality—at least, so says Descartes.

This is one of the most interesting subplots, and ironies, in the story of AI, because it was deductive logic, a field that Aristotle helped invent, that was the very first domino to fall.

Logic Gates

It begins, you might say, in the nineteenth century, when the English mathematician and philosopher George Boole works out and publishes a system for describing logic in terms of conjunctions of three basic operations: AND, OR,[10] and NOT. The idea is that you begin with any number of simple statements, and by passing them through

---

drunk.' By contrast, when it remains 'itself by itself' and investigates intelligibles, its straying comes to an end, and it achieves stability and wisdom."

10. The word "or" in English is ambiguous—"Do you want sugar or cream with your coffee?" and "Do you want fries or salad with your burger?" are actually two *different* types of questions. (In the first, "Yes"—meaning "both"—and "No"—meaning "neither"—are perfectly suitable answers, but in the second it's understood that you will choose one and exactly one of the options.) We respond differently, and appropriately, to each without often consciously noticing the difference. Logicians, to be more precise, use the terms "inclusive or" and "exclusive or" for these two types of questions, respectively. In Boolean logic, "OR" refers to the *inclusive* or, which means "either one or the other, *or both.*" The exclusive or—"either one or the other, *but not both*"—is written "XOR."

a kind of flowchart of ANDs, ORs, and NOTs, you can build up and break down statements of essentially endless complexity. For the most part, Boole's system is ignored, read only by academic logicians and considered of little practical use, until in the mid-1930s an undergraduate at the University of Michigan by the name of Claude Shannon runs into Boole's ideas in a logic course, en route to a mathematics and electrical engineering dual degree. In 1937, as a twenty-one-year-old graduate student at MIT, something clicks in his mind; the two disciplines bridge and merge like a deck of cards. You can implement Boolean logic *electrically,* he realizes, and in what has been called "the most important master's thesis of all time," he explains how. Thus is born the electronic "logic gate"—and soon enough, the processor.

Shannon notes, also, that you might be able to think of *numbers* in terms of Boolean logic, namely, by thinking of each number as a series of true-or-false statements about the numbers that it contains—specifically, which powers of 2 (1, 2, 4, 8, 16 . . .) it contains, because every integer can be made from adding up at most one of each. For instance, 3 contains 1 and 2 but not 4, 8, 16, and so on; 5 contains 4 and 1 but not 2; and 15 contains 1, 2, 4, and 8. Thus a set of Boolean logic gates could treat them as bundles of logic, true and false, yeses and noes. This system of representing numbers is familiar to even those of us who have never heard of Shannon or Boole—it is, of course, binary.

Thus, in one fell swoop, the master's thesis of twenty-one-year-old Claude Shannon will break the ground for the processor *and* for digital mathematics. And it will make his future wife's profession—although he hasn't met her yet—obsolete.

And it does more than that. It forms a major part of the recent history—from the mechanical logic gates of Charles Babbage through the integrated circuits in our computers today—that ends up amounting to a huge blow to humans' unique claim to and dominance of the area of "reasoning." Computers, lacking almost everything else that makes humans humans, have our *unique* piece in spades. They have more of it than we do. So what do we make of this? How has this affected and been affected by our sense of self? How *should* it?

First, let's have a closer look at the philosophy surrounding, and migration of, the self in times a little closer to home: the twentieth century.

Death Goes to the Head

Like our reprimanded ogler, like philosophy between Aristotle and Descartes, the gaze (if you will) of the medical community and the legal community moves upward too, abandoning the cardiopulmonary region as the brain becomes the center not only of life but of death. For most of human history, breath and heartbeat were the factors considered relevant for determining if a person was "dead" or not. But in the twentieth century, the determination of death became less and less clear, and so did its *definition*, which seemed to have less and less to do with the heart and lungs. This shift was brought on both by the rapidly increasing medical understanding of the brain, and by the newfound ability to restart and/or sustain the cardiopulmonary system through CPR, defibrillators, respirators, and pacemakers. Along with these changes, the increasing viability of organ donation added an interesting pressure to the debate: to declare certain people with a breath and a pulse "dead," and thus available for organ donation, could save the lives of others.[11] The "President's Commission for the Study of Ethical Problems in Medicine and Biomedical and Behavioral Research" presented Ronald Reagan in the summer of 1981 with a 177-page report called "Defining Death" wherein the American legal definition of death would be expanded, following the decision in 1968 of an ad hoc committee of the Harvard Medical School to *include* those with cardiopulmonary function (be it artificial or natural) who had sufficiently irreparable and severe brain

---

11. The heart needs the brain just as much as the brain needs the heart. But the heart—with all due respect—is fungible.

damage. The Uniform Determination of Death Act, passed in 1981, specifies "irreversible cessation of all functions of the entire brain, including the brain stem."

Our legal and medical definitions of death—like our sense of what it means to live—move to the brain. We look for death where we look for life.

The bulk of this definitional shift is by now long over, but certain nuances and more-than-nuances remain. For instance: Will damage to certain specific *areas* of the brain be enough to count? If so, which areas? The Uniform Determination of Death Act explicitly side-stepped questions of "neocortical death" and "persistent vegetative state"—questions that, remaining unanswered, have left huge medical, legal, and philosophical problems in their wake, as evidenced by the nearly decade-long legal controversy over Terri Schiavo (in a sense, over whether or not Terri Schiavo was legally "alive").

It's not my intention here to get into the whole legal and ethical and neurological scrum over death, per se—nor to get into the theological one about where exactly the soul-to-body downlink has been thought to take place. Nor to get into the metaphysical one about Cartesian "dualism"—the question of whether "mental events" and "physical events" are made up of one and the same, or two different, kinds of stuff. Those questions go deep, and they take us too far off our course. The question that interests me is how this anatomical shift affects and is affected by our sense of what it *means* to be alive and to be human.

That core, that essence, that meaning, seems to have migrated in the past few millennia, from the whole body to the organs in the chest (heart, lungs, liver, stomach) to the one in the head. Where next?

Consider, for instance, the example of the left and right hemispheres.

The human brain is composed of two distinct "cerebral hemispheres" or "half brains": the left and the right. These hemispheres communicate via an extremely "high bandwidth" "cable"—a bundle of roughly 200 million axons called the corpus callosum. With the exception of the data being ferried back and forth across the corpus callosum, the two halves of our brain operate independently—and rather differently.

The Split Brain

So: *where are we?*

Nowhere is this question raised more shockingly and eerily than in the case of so-called "split-brain" patients, whose hemispheres—usually as a result of surgical procedures aimed at reducing seizures—have been separated and can no longer communicate. "Joe," a split-brain patient, says, "You know, the left hemisphere and right hemisphere, now, are working independent of each other. But, you don't notice it . . . It doesn't feel any different than it did before."

It's worth considering that the "you"—in this case, a rhetorical stand-in for "I"—no longer applies to Joe's entire brain; the domain of that pronoun has shrunk. It only, now, refers to the left hemisphere, which happens to be the dominant hemisphere in language. Only that half, you might say, is speaking.

At any rate, Joe is telling us that "he"—or, his left hemisphere—doesn't notice anything different. But things *are* different, says his doctor, Michael Gazzaniga. "What we can do is play tricks, by putting information into his disconnected, mute, non-talking right hemisphere, and watch it produce behaviors. And out of that, we can really see that there is, in fact, reason to believe that there's all kinds of complex processes going on outside of his conscious awareness of his left half-brain."

In one of the more eerie experiments, Gazzaniga flashes two images—a hammer and a saw—to different parts of Joe's visual field, such that the hammer image goes to his left hemisphere and the saw to his right. "What'd you see?" asks Gazzaniga.

"I saw a hammer," Joe says.

Gazzaniga pauses. "So, just close your eyes, and draw with your left hand." Joe picks up a marker with his left hand, which is controlled by his right hemisphere. "Just let it go," says Gazzaniga. Joe's left hand draws a saw.

"That looks nice," says Gazzaniga. "What's that?"

"Saw?" Joe says, slightly confused.

"Yeah. What'd you see?"

"Hammer."

"What'd you draw that for?"

"I dunno," Joe, or at any rate his left hemisphere, says.

In another experiment, Gazzaniga flashes a chicken claw to a split-brain patient's "talking" left hemisphere and a snowbank to the "mute" right hemisphere. The patient draws a snow shovel, and when Gazzaniga asks him why he drew a shovel, he doesn't hesitate or shrug. Without missing a beat, he says, "Oh, that's simple. The chicken claw goes with the chicken, and you need a shovel to clean out the chicken shed." Of course, this as an explanation is completely false.

The left hemisphere, it seems, is constantly drawing cause-and-effect inferences from experience, constantly attempting to make sense of events. Gazzaniga dubs this module, or whatever it is exactly, the "interpreter." The interpreter, split-brain patients show us, has no problem and no hesitation confabulating a false causation or a false motive. Actually, "lying" is putting it too strongly—it's more like "confidently asserting its best guess." Without access to what's happening in the right hemisphere, that guess can sometimes be purely speculative, as in this case. What's fascinating, though, is that this interpreter doesn't necessarily even get it right all the time in a *healthy* brain.

To take a random example: a woman undergoing a medical procedure had her "supplementary motor cortex" stimulated electrically, producing uncontrollable laughter. But instead of being bewildered by this inexplicable outburst, she acted as if anyone in her position would have cracked up: "You guys are just so *funny* standing around!"

I find it so tragic that when an infant cries, the parents sometimes have no idea what might be the cause of the cry—hunger? thirst? dirty diaper? fatigue? If only the child could *tell* them! But no, they must simply run down the list—here's some food, no, still crying, here's a new diaper, no, still crying, here's your blanket, maybe you need a nap, no, still crying . . . But it occurs to me that this also describes my relationship to *myself*. When I am in a foul mood, I think, "How's my work going? How's my social life? How's my love life? How much water have I had today? How much coffee have I had today? How well have I been eating? How much have I been exercising? How have I been sleeping? How's the weather?" And sometimes that's the best I can do: eat some fruit, jog around the neighborhood, take a nap, and on and on till the mood changes and I think, "Oh, I guess that was it." I'm not much better than the infant.

Once, in graduate school, after making what I thought was not a trivial, but not a particularly major, life decision, I started to feel kind of "off." The more off I felt, the more I started to rethink my decision, and the more I rethought the decision—this was on the bus on the way to campus—the more I started feeling nauseous, sweaty, my blood running hot and cold. "Oh my God!" I remember thinking. "This is actually a much bigger deal than I thought!" No, it was simply that I'd caught the stomach flu going around the department that month.

You see this happen—"misattribution"—in all sorts of fascinating studies. For instance: they've proven that people look more attractive to you when you're walking across a suspension bridge or riding a roller coaster. Apparently, the body generates all this jitteriness, which is actually fear, but the rational mind says something to the effect of, "Oh, butterflies in the stomach! But obviously there's nothing to be afraid of from a silly little roller coaster or bridge—they're completely safe. So it must be the person standing next to me that's got me all aflutter . . ." In a Canadian study, a woman gave her number to male hikers either just before they reached the Capilano Suspension Bridge or in the middle of the bridge. Those who met her *on* the bridge were twice as likely to call and ask for a date.

Someone who can put together extremely compelling reasons for why they did something can get themselves out of hot water more often than someone at a loss for why they did it. But just because a person gives you a sensible explanation for a strange or objectionable behavior, *and* the person is an honest person, doesn't mean the explanation is correct. And the ability to spackle something plausible into the gap between cause and effect doesn't make the person any more rational, or responsible, or moral, even though we'll pretty consistently judge them so.

Says Gazzaniga, "What Joe, and patients like him, and there are many of them, teaches us, is that the mind is made up of a constellation of independent, semi-independent, agents. And that these agents, these processes, carry on a vast number of activities outside of our conscious awareness."

"Our conscious awareness"—*our*! The implication here (which Gazzaniga later confirms explicitly) is that Joe's "I" pronoun may have *always* referred mostly, and primarily, to his left hemisphere. So, he says, do ours.

Hemispheric Chauvinism: Computer and Creature

"The entire history of neurology and neuropsychology can be seen as a history of the investigation of the left hemisphere," says neurologist Oliver Sacks.

One important reason for the neglect of the right, or "minor," hemisphere, as it has always been called, is that while it is easy to demonstrate the effects of variously located lesions on the left side, the corresponding syndromes of the right hemisphere are much less distinct. It was presumed, usually contemptuously, to be more "primitive" than the left, the latter being seen as the unique flower of human evolution. And in a sense this is correct: the left hemisphere is more sophisticated and specialised, a very late outgrowth of the primate, and especially the hominid, brain. On the other hand, it is the right hemisphere which controls the crucial powers of recognising reality which every living creature must have in order to survive. The left hemisphere, like a computer tacked onto the basic creatural brain, is designed for programs and schematics; and classical neurology was more concerned with schematics than with reality, so that when, at last, some of the right-hemisphere syndromes emerged, they were considered bizarre.

The neurologist V. S. Ramachandran echoes this sentiment:

The left hemisphere is specialized not only for the actual production of speech sounds but also for the imposition of syntactic structure on speech and for much of what is called semantics—comprehension of meaning. The right hemisphere, on the other hand, doesn't govern spoken words but seems to be concerned with more subtle aspects of language such as nuances of metaphor, allegory and ambiguity—skills that are inadequately emphasized in our elementary schools but that are vital for the advance of civilizations through poetry, myth and drama. We tend to call the left hemisphere the major or "dominant" hemisphere because it, like a chauvinist, does all the talking (and maybe much of the internal thinking as well), claiming to be the repository of humanity's highest attribute, language.

"Unfortunately," he explains, "the mute right hemisphere can do nothing to protest."

Slightly to One Side

This odd focus on, and "dominance" of, the left hemisphere, says arts and education expert (and knight) Sir Ken Robinson, is evident in the hierarchy of subjects within virtually all of the world's education systems:

> At the top are mathematics and languages, then the humanities, and the bottom are the arts. Everywhere on Earth. And in pretty much every system too, there's a hierarchy within the arts. Art and music are normally given a higher status in schools than drama and dance. There isn't an education system on the planet that teaches dance every day to children the way we teach them mathematics. Why? Why not? I think this is rather important. I think math is very important, but so is dance. Children dance all the time if they're allowed to; we all do. We all have bodies, don't we? Did I miss a meeting? Truthfully, what happens is, as children grow up, we start to educate them progressively from the waist up. And then we focus on their heads. And slightly to one side.

That side, of course, being the left.

The American school system "promotes a catastrophically narrow idea of intelligence and ability," says Robinson. If the left hemisphere, as Sacks puts it, is "like a computer tacked onto the basic creatural brain," then by identifying ourselves with the goings-on of the left hemisphere, by priding ourselves on it and "locating" ourselves in it, we start to regard ourselves, in a manner of speaking, as computers. By better educating the left hemisphere and better valuing and rewarding and nurturing its abilities, we've actually started *becoming* computers.

Rational Agents

You see the same left-hemisphere bias in the field of economics. Emotions are considered barnacles on the smooth hull of the mind. Decisions should be made, to the greatest extent possible, in their absence—and, as much as possible, calculatingly, even algorithmically.

"If you had asked Benjamin Franklin, 'How should I go about making up my mind?' " says Baba Shiv of the Stanford Graduate School of Business, "what he would have advised you to do is, list down all of the positives and all the negatives of your present option, list down all of the positives and all the negatives of the alternative that you have. And then choose that option that has got the greatest number of positives and the least number of negatives."

This analytical, emotionless notion of ideal decision making was codified into the "rational agent" model of economic theory. The model consumer or investor, it reckoned, would somehow have access to all possible information about the market and would be able to somehow instantly distill it all and make the perfect choice. Shockingly, real markets, and real investors and consumers, don't work this way.

But even when recognition came that omniscient rationality was *not* the right model to use, it seemed that economists were more interested in talking about this as a shortfall than a boon. Consider 2008's *Predictably Irrational,* in which behavioral economist Dan Ariely argues against the rational-agent model by highlighting the various human behaviors that don't accord with it. A victory for re-assimilating the various neglected and denigrated capacities of the self? A glance at the jacket blurbs is enough to produce a resounding no, revealing the light in which we are meant to read these deviations from economic theory. "How we can prevent being fooled," says Jerome Groopman, Recanati Professor of Medicine at Harvard Medical School. "The weird ways we act," says business writer James Surowiecki. "Foibles, errors, and bloopers," says Harvard psychologist Daniel Gilbert. "Foolish, and sometimes disastrous, mistakes," says Nobel laureate in economics George Akerlof. "Managing your emotions . . . so challenging for all of us . . . can help you avoid common mistakes," says financial icon Charles Schwab.[12]

---

12. The book's sequel, *The Upside of Irrationality,* is much more sanguine about "irrationality" in its title, if somewhat less so in the text itself.

Now, some of what passes for "irrationality" in traditional "rational" economics is simply bad science, cautions Daniel Kahneman, Nobel laureate from Princeton. For instance, given a choice between a million dollars and a 50 percent chance of winning four million dollars, the "rational" choice is "obviously" the latter, whose "expected outcome" is two million dollars, double the first offer. Yet most people say they would choose the former—fools! Or are they? It turns out to depend on how wealthy you are: the richer you are, the more inclined toward the gamble. Is this because wealthier people are (as demonstrated by being rich) more logical? Is this because less wealthy people are blinded by an emotional reaction to money? Is it because the brain is, tragically, more averse to loss than excited by gain? Or perhaps the wealthy person who accepts the gamble and the less wealthy person who declines it are, in fact, choosing completely appropriately in both cases. Consider: a family deep into debt and about to default on their home could *really* use that first million; the added three million would be icing on the cake but wouldn't change much. The "quadruple or nothing" offer just isn't worth betting the farm—literally. Whereas for a billionaire like Donald Trump, a million bucks is chump change, and he'll probably take his chances, knowing the odds favor him. The two choose differently—and both choose *correctly*.

At any rate, and with examples like this one aside, the prevailing attitude seems clear: economists who subscribe to the rational-choice theory and those who critique it (in favor of what's known as "bounded rationality") *both* think that an emotionless, Spock-like approach to decision making is demonstrably superior. We should all aspire to throw off our ape ancestry to whatever extent we can—alas, we are fallible and will still make silly emotion-tinged "bloopers" here and there.

This has been for centuries, and by and large continues to be, the theoretical mainstream, and not just economics but Western intellectual history at large is full of examples of the creature needing the computer. But examples of the *reverse*, of the computer needing the creature, have been much rarer and more marginal—until lately.

Baba Shiv says that as early as the 1960s and '70s, evolutionary biologists began to ask—well, if the emotional contribution to decision making is so terrible and detrimental, why did it develop? If it was so bad, wouldn't we have evolved differently? The rational-choice theorists, I imagine, would respond by saying something like "we're on our way there, but just not fast enough." In the late '80s and through the '90s, says Shiv, neuroscientists "started providing evidence for the diametric opposite viewpoint" to rational-choice theory: "that emotion is *essential* for and *fundamental* to making good decisions."

Shiv recalls a patient he worked with "who had an area of the emotional brain knocked off" by a stroke. After a day of doing some tests and diagnostics for which the patient had volunteered, Shiv offered him a free item as a way of saying "thank you"—in this case, a choice between a pen and a wallet. "If you're faced with such a trivial decision, you're going to examine the pen, examine the wallet, think a little bit, grab one, and go," he says. "That's it. It's non-consequential. It's just a pen and a wallet. This patient didn't do that. He does the same thing that we would do, examine them and think a little bit, and he grabs the pen, starts walking—hesitates, grabs the wallet. He goes outside our office—comes back and grabs the pen. He goes to his hotel room—believe me: inconsequential a decision!—he leaves a message on our voice-mail mailbox, saying, 'When I come tomorrow, can I pick up the wallet?' This constant state of indecision."

USC professor and neurologist Antoine Bechara had a similar patient, who, needing to sign a document, waffled between the two pens on the table for a full twenty minutes.[13] (If we are some

---

13. Neurologist Antonio Damasio showed him a series of extremely emotionally charged pictures—a severed foot, a naked woman, a burning home—to which he barely reacted. Fans of *Blade Runner* or Philip K. Dick will recall this as almost the spitting image of the fictitious "Voigt-Kampff test." Good thing he didn't live in the *Blade Runner* universe: Harrison Ford would have decided this man was a "replicant"—and killed him.

computer/creature hybrid, then it seems that damage to the creature-type impulses and forces may leave us vulnerable to computer-type problems, like freezing and looping.) In cases like this there is no "rational" or "correct" answer. So the logical, analytical mind just flounders and flounders.

In other decisions where there is no objectively best choice, where there are simply a number of subjective variables with trade-offs between them (airline tickets is one example, houses another, and Shiv includes "mate selection"—a.k.a. dating—among these), the hyperrational mind basically freaks out, something that Shiv calls a "decision dilemma." The nature of the situation is such that additional information probably won't even help. In these cases—consider the parable of the donkey that, halfway between two bales of hay and unable to decide which way to walk, starves to death—what we want, more than to be "correct," is to be *satisfied* with our choice (and out of the dilemma).

Shiv practices what he preaches. His and his wife's marriage was arranged—they decided to tie the knot after talking for twenty minutes[14]—and they committed to buying their house at first sight.

Coming Back to Our Senses

All this "hemispheric bias," you might call it, or rationality bias, or analytical bias—for it's in actuality more about analytical thought and linguistic articulation than about the left hemisphere *per se*—both compounds and is compounded by a whole host of other prevailing societal winds to produce some decidedly troubling outcomes.

I think back, for instance, to my youthful days in CCD—Confraternity of Christian Doctrine, or Catholicism night classes for kids in secular public schools. The ideal of piousness, it seemed to me in those days, was the life of a cloistered monk, attempting a kind of afterlife on earth by living, as much as possible, apart from the "creatural"

---

14. The *ultimate* Turing test victory, you might say.

aspects of life. The Aristotelian ideal: a life spent entirely in contemplation. No rich foods, no aestheticizing the body with fashion, no reveling in the body qua body through athletics—nor dancing—nor, of course, sex. On occasion making music, yes, but music so beholden to prescribed rules of composition and to mathematical ratios of harmony that it too seemed to aspire toward pure analytics and detachment from the general filth and fuzziness of embodiment.

And so for many of my early years I distrusted my body, and all the weird feelings that came with it. I *was* a mind, but merely *had* a body—whose main purpose, it seemed, was to move the mind around and otherwise only ever got in its way. I was consciousness—in Yeats's unforgettable words—"sick with desire / And fastened to a dying animal." After that animal finally did die, it was explained to me, things would get a lot better. They then made sure to emphasize that suicide is strictly against the rules. We were all in this thing together, and we all just had to wait this embodiment thing out.

Meanwhile, on the playground, I was contemptuous of the seemingly Neanderthal boys who shot hoops and grunted their way through recess—meanwhile, my friends and I talked about MS-DOS and Stephen Hawking. I tended to view the need to eat as an annoyance—I'd put food in my mouth to hush my demanding stomach the way a parent gives a needy infant a pacifier. Eating was *annoying*; it got in the way of *life*. Peeing was annoying, showering was annoying, brushing the crud off my teeth every morning and night was annoying, sleeping a third of my life away was annoying. And sexual desire—somehow I'd developed the idea that my first boyhood forays into masturbation had stamped my one-way ticket to hell—sexual desire was so annoying that I was pretty sure it had already cost me everything.

I want to argue that this Aristotelian/Stoic/Cartesian/Christian emphasis on reason, on thought, on the head, this distrust of the senses, of the body, has led to some *profoundly* strange behavior—and not just in philosophers, lawyers, economists, neurologists, educators, and the hapless would-be pious, but seemingly everywhere. In a world of manual outdoor labor, the sedentary and ever-feasting nobility made a status symbol of being overweight and pale; in a world of information work, it is a luxury to be tan and lean, if artificially or unhealthily so. Both scenarios would seem less than ideal. The very fact that we, as a rule, must deliberately "get exercise" bodes poorly: I imagine the middle-class city dweller paying money for a parking space or transit pass in lieu of walking a mile or two to the office, who then pays more money for a gym membership (and drives or buses there). I grew up three miles from the Atlantic Ocean; during the summer, tanning salons a block and a half from the beach would still be doing a brisk business. To see ourselves as distinct and apart from our fellow creatures is to see ourselves as distinct and apart from our *bodies*. The results of adopting this philosophy have been rather demonstrably weird.

Turing Machines and the Corporeal IOU

Wanting to get a handle on how these questions of soul and body intersect computer science, I called up the University of New Mexico's and the Santa Fe Institute's Dave Ackley, a professor in the field of artificial life.

"To me," he says, "and this is one of the rants that I've been on, that ever since von Neumann and Turing and the ENIAC guys[15] built machines, the model that they've used is the model of the conscious mind—one thing at a time, nothing changing except by conscious thought—no interrupts, no communication from the outside world. So in particular the computation was not only unaware of the world; it didn't realize that it had a body, so the computation was disembod-

---

15. John Mauchly and J. Presper Eckert, of the University of Pennsylvania. ENIAC (Electronic Numerical Integrator and Computer), built in 1946 and initially used in the calculations for the hydrogen bomb, was the first fully electronic and fully general-purpose computing machine.

ied, in a very real and literal sense. There's this IOU for a body that we wrote to computers ever since we designed them, and we haven't really paid it off yet."

I end up wondering if we even set out to *owe* computers a body. With the Platonic/Cartesian ideal of sensory mistrust, it seems almost as if computers were designed with the intention of *our* becoming more like *them*—in other words, computers represent an IOU of disembodiment that we wrote to ourselves. Indeed, certain schools of thought seem to imagine computing as a kind of oncoming rapture. Ray Kurzweil (in 2005's *The Singularity Is Near*), among several other computer scientists, speaks of a utopian future where we shed our bodies and upload our minds into computers and live forever, virtual, immortal, disembodied. Heaven for hackers.

To Ackley's point, most work on computation has not traditionally been on dynamic systems, or interactive ones, or ones integrating data from the real world in real time. Indeed, theoretical models of the computer—the Turing machine, the von Neumann architecture—seem like reproductions of an idealized version of conscious, deliberate reasoning. As Ackley puts it, "The von Neumann machine is an image of one's conscious mind where you tend to think: you're doing long division, and you run this algorithm step-by-step. And that's not how brains operate. And only in various circumstances is that how *minds* operate."

I spoke next with University of Massachusetts theoretical computer scientist Hava Siegelmann, who agreed. "Turing was very [mathematically] smart, and he suggested the Turing machine as a way to describe a *mathematician*.[16] It's [modeling] the way a person solves a problem, not the way he recognizes his mother." (Which latter problem, as Sacks suggests, is of the "right hemisphere" variety.)

For some time in eighteenth-century Europe, there was a sweep-

---

16. Recall Turing: "The idea behind digital computers may be explained by saying that these machines are intended to carry out any operations which could be done by a human computer."

ing fad of automatons: contraptions made to look and act as much like real people or animals as possible. The most famous and celebrated of these was the "Canard Digérateur"—the "Digesting Duck"—created by Jacques de Vaucanson in 1739. The duck provoked such a sensation that Voltaire himself wrote of it, albeit with tongue in cheek: "Sans . . . le canard de Vaucanson vous n'auriez rien qui fit ressouvenir de la gloire de la France," sometimes humorously translated as "Without the shitting duck we'd have nothing to remind us of the glory of France."

Actually, despite Vaucanson's claims that he had a "chemistry lab" inside the duck mimicking digestion, there was simply a pouch of bread crumbs, dyed green, stashed behind the anus, to be released shortly after eating. Stanford professor Jessica Riskin speculates that the lack of attempt to simulate digestion had to do with a feeling at the time that the "clean" processes of the body could be mimicked (muscle, bone, joint) with gears and levers but that the "messy" processes (mastication, digestion, defecation) could not. Is it possible that something similar happened in our approach to mimicking the mind?

In fact, the field of computer science split, very early on, between researchers who wanted to pursue more "clean," algorithmic types of structures and those who wanted to pursue more "messy" and gestalt-oriented structures. Though both have made progress, the "algorithmic" side of the field has, from Turing on, completely dominated the more "statistical" side. That is, until recently.

Translation

There's been interest in neural networks and analog computation and more statistical, as opposed to algorithmic, computing since at least the early 1940s, but the dominant paradigm by far was the algorithmic, rule-based paradigm—that is, up until about the turn of the century.

If you isolate a specific type of problem—say, the problem of machine translation—you see the narrative clear as day. Early approaches were about building huge "dictionaries" of word-to-word pairings, based on meaning, and algorithms for turning one syntax and grammar into another (e.g., if going to Spanish from English, move the adjectives that come before a noun so that they come after it).

To get a little more of the story, I spoke on the phone with computational linguist Roger Levy of UCSD. Related to the problem of translation is the problem of paraphrase. "Frankly," he says, "as a computational linguist, I can't imagine trying to write a program to pass the Turing test. Something I might do as a confederate is to take a sentence, a relatively complex sentence, and say, 'You said this. You could also express the meaning with this, this, this, and this.' That would be extremely difficult, paraphrase, for a computer." But, he explains, such specific "demonstrations" on my part might backfire: they come off as unnatural, and I might have to explicitly lay out a case for why what I'm saying is hard for a computer to do. "All this depends on the informedness level of the judge," he says. "The nice thing about small talk, though, is that when you're in the realm of heavy reliance on pragmatic inferences, that's very hard for a computer—because you have to rely on real-world knowledge."

I ask him for some examples of how "pragmatic inferences" might work. "Recently we did an experiment in real-time human sentence comprehension. I'm going to give you an ambiguous sentence: 'John babysat the child of the musician, who is arrogant and rude.' Who's rude?" I said that to my mind it's the musician. "Okay, now: 'John detested the child of the musician, who is arrogant and rude.' " Now it sounds like the child is rude, I said. "Right. No system in existence has this kind of representation."

It turns out that all kinds of everyday sentences require more than just a dictionary and a knowledge of grammar—compare "Take the pizza out of the oven and then close it" with "Take the pizza out of the oven and then put it on the counter." To make sense of the pronoun "it" in these examples, and in ones like "I was holding the coffee cup and the milk carton, and just poured it in without checking the expiration date," requires an understanding of how the *world* works, not how the *language* works. (Even a system programmed with basic facts like "coffee and milk are liquids," "cups and cartons are containers," "only liquids can be 'poured,' " etc., won't be able to tell whether pouring the coffee into the carton or the milk into the cup makes more sense.)

A number of researchers feel that the attempt to break language down with thesauri and grammatical rules is simply not going to crack the translation problem. A new approach abandons these strategies, more or less entirely. For instance, the 2006 NIST machine translation competition was convincingly won by a team from Google, stunning a number of machine translation experts: not a single human on the Google team knew the languages (Arabic and Chinese) used in the competition. And, you might say, neither did the software itself, which didn't give a whit about meaning or about grammar rules. It simply drew from a massive database of high-quality human translation[17] (mostly from the United Nations minutes, which are proving to be the twenty-first century's digital Rosetta stone), and patched phrases together according to what had been done in the past. Five years later, these kinds of "statistical" techniques are still imperfect, but they have left the rule-based systems pretty firmly in the dust.

Among the other problems in which statistical, as opposed to rule-based, systems are triumphing? One of our right-hemisphere paragons: object recognition.

UX

Another place where we're seeing the left-hemisphere, totally deliberative, analytical approach erode is with respect to a concept called

---

17. Interestingly, this means that paraphrasing is actually *harder* for computers than translation, because there aren't huge paraphrase corpora lying around ready to become statistical fodder. The only examples I can think of off the top of my head would be, ironically, competing translations: of famous works of literature and religious texts.

UX, short for User Experience—it refers to the experience a given user has using a piece of software or technology, rather than the purely *technical* capacities of that device. The beginnings of computer science were dominated by concerns for the technical capacities, and the exponential growth in processing power[18] during the twentieth century made the 1990s, for instance, an exciting time. Still, it wasn't a *beautiful* time. My schoolmate brought us over to show us the new machine he'd bought—it kept overheating, so he'd opened the case up and let the processor and motherboard dangle off the edge of the table by the wires, where he'd set up his room fan to blow the hot air out the window. The keyboard keys stuck when you pressed them. The mouse required a cramped, *T. rex*–claw grip. The monitor was small and tinted its colors slightly. But computationally, the thing could scream.

This seemed the prevailing aesthetic of the day. My first summer job, in eighth grade—rejected as a busboy at the diner, rejected as a caddy at the golf course, rejected as a summer camp counselor—was at a web design firm, where I was the youngest employee by at least a decade, and the lowest paid by a factor of 500 percent, and where my responsibilities in a given day would range from "Brian, why don't you restock the toilet paper and paper towels in the bathrooms" to "Brian, why don't you perform some security testing on the new e-commerce intranet platform for Canon." I remember my mentor figure at the web design company saying, in no uncertain terms, "function over form."

The industry as a whole seemed to take this mantra so far that function began trumping *function*: for a while, an arms race between hardware and software created the odd situation that computers were getting exponentially faster but no faster at all to *use*, as software made ever-larger demands on system resources, at a rate that

---

18. This trend is described by what's called Moore's Law, the 1965 prediction of Intel's co-founder Gordon Moore that the number of transistors in a processor would double every two years.

matched and sometimes outpaced hardware improvements. (For instance, Office 2007 running on Windows Vista uses twelve times as much memory and three times as much processing power as Office 2000 running on Windows 2000, with nearly twice as many execution threads as the immediately previous version.) This is sometimes called "Andy and Bill's Law," referring to Andy Grove of Intel and Bill Gates of Microsoft: "What Andy giveth, Bill taketh away." Users were being subjected to the very same lags and lurches on their new machines, despite exponentially increasing computing power, all of which was getting sopped up by new "features." Two massive companies pouring untold billions of dollars and thousands of man-years into advancing the cutting edge of hardware and software, yet the advances effectively canceled out. The user experience went nowhere.

I think we're just in the past few years seeing the consumer and corporate attitude changing. Apple's first product, the Apple I, did not include a keyboard or a monitor—it didn't even include a *case* to hold the circuit boards. But it wasn't long before they began to position themselves as prioritizing user experience ahead of power—and ahead of pricing. Now they're known, by admirers and deriders alike, for machines that manage something which seemed either impossible or irrelevant, or both, until a few years ago—elegance.

Likewise, as computing technology moves increasingly toward mobile devices, product development becomes less about the raw computing horsepower and more about the overall design of the product and its fluidity, reactivity, and ease of use. This fascinating shift in computing emphasis may be the cause, effect, or correlative of a healthier view of human intelligence—not so much that it is complex and powerful, per se, as that it is reactive, responsive, sensitive, nimble. The computers of the twentieth century helped us to see that.

Centering Ourselves

We are computer tacked onto creature, as Sacks puts it. And the point isn't to denigrate one, or the other, any more than a catamaran ought to become a canoe. The point isn't that we're half lifted out of beastliness by reason and can try to get even further through force of will. The tension is the point. Or, perhaps to put it better, the collaboration, the dialogue, the duet.

The word games Scattergories and Boggle are played differently but scored the same way. Players, each with a list of words they've come up with, compare lists and cross off every word that appears on more than one list. The player with the most words remaining on her sheet wins. I've always fancied this a rather cruel way of keeping score. Imagine a player who comes up with four words, and each of her four opponents only comes up with one of them. The round is a draw, but it hardly feels like one . . . As the line of human uniqueness pulls back ever more, we put the eggs of our identity into fewer and fewer baskets; then the computer comes along and takes that final basket, crosses off that final word. And we realize that uniqueness, per se, never had anything to do with it. The ramparts we built to keep other species and other mechanisms out also kept us in. In breaking down that last door, computers have let us out. And back into the light.

Who would have imagined that the computer's *earliest* achievements would be in the domain of logical analysis, a capacity held to be what made us most different from everything on the planet? That it could drive a car and guide a missile before it could ride a bike? That it could make plausible preludes in the style of Bach before it could make plausible small talk? That it could translate before it could paraphrase? That it could spin half-plausible postmodern theory essays[19] before it could be shown a chair and say, as any toddler can, "chair"?

---

19. "If one examines capitalist discourse, one is faced with a choice: either reject nihilism or conclude that the goal of the writer is social comment, given that the premise of Foucaultist power relations is valid." Or, "Thus, the subject is interpolated into a nihilism that includes consciousness as a paradox." Two sentences of infinitely many at www.elsewhere.org/pomo.

We forget what the impressive things are. Computers are reminding us.

One of my best friends was a barista in high school: over the course of the day she would make countless subtle adjustments to the espresso being made, to account for everything from the freshness of the beans to the temperature of the machine to the barometric pressure's effect on the steam volume, meanwhile manipulating the machine with octopus-like dexterity and bantering with all manner of customers on whatever topics came up. Then she goes to college, and lands her first "real" job—rigidly procedural data entry. She thinks longingly back to her barista days—a job that actually made demands of her intelligence.

I think the odd fetishization of analytical thinking, and the concomitant denigration of the creatural—that is, animal—and embodied aspect of life is something we'd do well to leave behind. Perhaps we are finally, in the beginnings of an age of AI, starting to be able to *center* ourselves again, after generations of living "slightly to one side."

Besides, we know, in our capitalist workforce and precapitalist-workforce education system, that specialization and differentiation are important. There are countless examples, but I think, for instance, of the 2005 book *Blue Ocean Strategy: How to Create Uncontested Market Space and Make the Competition Irrelevant,* whose main idea is to avoid the bloody "red oceans" of strident competition and head for "blue oceans" of uncharted market territory. In a world of only humans and animals, biasing ourselves in favor of the left hemisphere might make some sense. But the arrival of computers on the scene changes that dramatically. The bluest waters aren't where they used to be.

Add to this that humans' contempt for "soulless" animals, their unwillingness to think of themselves as descended from their fellow "beasts," is now cut back on all kinds of fronts: growing secularism and empiricism, growing appreciation for the cognitive and behavioral abilities of organisms other than ourselves, and, not coincidentally, the entrance onto the scene of a being far more soulless than any common chimpanzee or bonobo—in this sense AI may even turn out to be a boon for animal rights.

Indeed, it's entirely possible that we've seen the high-water mark of the left-hemisphere bias. I think the return of a more balanced view of the brain and mind—and of human identity—is a good thing, one that brings with it a changing perspective on the sophistication of various tasks.

It's my belief that only experiencing and understanding *truly* disembodied cognition, only seeing the coldness and deadness and disconnectedness of something that truly *does* deal in pure abstraction, divorced from sensory reality, only this can snap us out of it. Only this can bring us, quite literally, back to our senses.

One of my graduate school advisers, poet Richard Kenney, describes poetry as "the mongrel art—speech on song," an art he likens to lichen: that organism which is actually not an organism at all but a cooperation between fungi and algae so common that the cooperation itself seemed a species. When, in 1867, the Swiss botanist Simon Schwendener first proposed the idea that lichen was in fact two organisms, Europe's leading lichenologists ridiculed him—including Finnish botanist William Nylander, who had taken to making allusions to "stultitia Schwendeneriana," fake botanist-Latin for "Schwendener the simpleton." Of course, Schwendener happened to be completely right. The lichen is an odd "species" to feel kinship with, but there's something fitting about it.

What appeals to me about this notion—the mongrel art, the lichen, the monkey and robot holding hands—is that it seems to describe the human condition too. Our very essence is a kind of mongrelism. It strikes me that some of the best and most human emotions come from this lichen state of computer/creature interface, the admixture, the estuary of desire and reason in a system aware enough to apprehend its own limits, and to push at them: curiosity, intrigue, enlightenment, wonder, awe.

Ramachandran: "One patient I saw—a neurologist from New York—suddenly at the age of sixty started experiencing epileptic seizures arising from his right temporal lobe. The seizures were alarming, of course, but to his amazement and delight he found himself becoming fascinated by poetry, for the first time in his life. In fact, he began thinking in verse, producing a voluminous outflow of rhyme. He said that such a poetic view gave him a new lease on life, a fresh start just when he was starting to feel a bit jaded."

Artificial intelligence may very well be such a seizure.

4. Site-Specificity vs. Pure Technique

Introducing ELIZA

SENT: Men are all alike.

RECEIVED: In what way?

SENT: They're always bugging us about something or other.

RECEIVED: Can you think of a specific example?

SENT: Well, my boyfriend made me come here.

RECEIVED: Your boyfriend made you come here.

SENT: He says I'm depressed much of the time.

RECEIVED: I am sorry to hear you are depressed.

SENT: It's true. I am unhappy.

RECEIVED: Do you think coming here will help you not to be unhappy?

SENT: I need some help, that much seems certain.

RECEIVED: What would it mean to you if you got some help?

SENT: Perhaps I could learn to get along with my mother.

RECEIVED: Tell me more about your family.

When the Turing test was first proposed in 1950, it was a hypothetical: technology was nowhere near a point at which a practical test was possible. But, as we know, it got there, with the first conversational computer program to attract significant notice and attention being

ELIZA,[1] written in 1964 and 1965 by Joseph Weizenbaum at MIT. The history of conversational computer programs contains every bit as many colorful "characters" in the programs themselves as it does in the humans that created them, and ELIZA's story is an interesting one. Modeled after a Rogerian therapist, ELIZA worked on a very simple principle: extract key words from the user's own language, and pose their statements back to them. ("I am __unhappy__." "Do you think coming here will help you not to be __unhappy__?") If in doubt, it might fall back on some completely generic phrases like "Please go on." This technique of fitting the user's statements into a set of predefined patterns and responding with a prescribed phrasing of its own—called "template matching"—was ELIZA's only capacity.

The results were stunning, maybe even staggering, considering that ELIZA was essentially the first chat program ever written, with essentially no memory, no processing power, and written in just a couple hundred lines of code: many of the people who first talked with ELIZA were convinced that they were having a genuine human interaction. In some cases even Weizenbaum's own insistence was of no use. People would ask to be left alone to talk "in private," sometimes for hours, and returned with reports of having had a meaningful therapeutic experience. Meanwhile, academics leaped to conclude that ELIZA represented "a general solution to the problem of computer understanding of natural language." Appalled and horrified, Weizenbaum did something almost unheard of: an immediate about-face of his entire career. He pulled the plug on the ELIZA

---

1. "Her" name is an allusion to Eliza Doolittle, the main character in George Bernard Shaw's 1913 play *Pygmalion*. Inspired by the myth of Pygmalion, a sculptor who creates a sculpture so realistic he falls in love with it (which also inspired, among many other works, *Pinocchio*), Shaw's play (itself the inspiration for the musical *My Fair Lady*) takes this idea and makes it into a tale of fluency and class: a phonetics professor makes a bet that he can train the lower-class Eliza Doolittle in the spoken English of the aristocracy and have her pass as a noble—a kind of Turing test in its own right. It's easy to see why Weizenbaum drew from Shaw's *Pygmalion* to name his therapist; unfortunately, he ended up with something closer to Ovid's story than Shaw's.

project, encouraged his own critics, and became one of science's most outspoken opponents of AI research.

But in some sense the genie was already out of the bottle, and there was no going back. The basic template-matching skeleton and approach of ELIZA have been reworked and implemented in some form or other in almost every chat program since, including the contenders at the Loebner Prize. And the enthusiasm, unease, and controversy surrounding these programs have only grown.

One of the strangest twists to the ELIZA story, however, was the reaction of the *medical* community, which, too, decided Weizenbaum had hit upon something both brilliant and useful with ELIZA. The *Journal of Nervous and Mental Disease,* for example, said of ELIZA in 1966: "If the method proves beneficial, then it would provide a therapeutic tool which can be made widely available to mental hospitals and psychiatric centers suffering a shortage of therapists. Because of the time-sharing capabilities of modern and future computers, several hundred patients an hour could be handled by a computer system designed for this purpose. The human therapist, involved in the design and operation of this system, would not be replaced, but would become a much more efficient man since his efforts would no longer be limited to the one-to-one patient-therapist ratio as now exists."

Famed scientist Carl Sagan, in 1975, concurred: "No such computer program is adequate for psychiatric use today, but the same can be remarked about some human psychotherapists. In a period when more and more people in our society seem to be in need of psychiatric counseling, and when time sharing of computers is widespread, I can imagine the development of a network of computer psychotherapeutic terminals, something like arrays of large telephone booths, in which, for a few dollars a session, we would be able to talk with an attentive, tested, and largely non-directive psychotherapist."

Incredibly, it wouldn't be long into the twenty-first century before this prediction—again, despite all possible protestation Weizenbaum could muster—came true. The United Kingdom's National Institute for Health and Clinical Excellence recommended in 2006 that cognitive-behavioral therapy software (which, in this case, doesn't pretend it's a human) be made available in England and Wales as an early treatment option for patients with mild depression.

Scaling Therapy

With ELIZA, we get into some serious, profound, even grave questions about psychology. Therapy is always *personal*. But does it actually need to be personal*ized*? The idea of having someone talk with a computerized therapist is not really all that much less intimate than having them read a book.[2] Take, for instance, the 1995 bestseller *Mind over Mood*: it's one-size-fits-all cognitive-behavioral therapy. Is such a thing appropriate?

(On Amazon, one reviewer lashes out against *Mind Over Mood*: "All experiences have meaning and are rooted in a context. There is not [sic] substitute for seeking the support of a well trained, sensitive psychotherapist before using such books to 'reprogram' yourself. Remember, you're a person, not a piece of computer software!" Still, for every comment like this, there are about thirty-five people saying that just following the steps outlined in the book changed their lives.)

There's a Sting lyric in "All This Time" that's always broken my heart: "Men go crazy in congregations / They only get better one by one." Contemporary women, for instance, are all dunked into the same mass-media dye bath of body-image issues, and then each, indi-

---

2. The only difference, which may be an important one, is that the book's boundaries and *extents* are clear. If you read it front to back, you know exactly which areas it covers and which it doesn't. The bot's extents are less clear: you must, probing the bot verbally, find them. One can imagine a bot that contains a useful response that the user simply doesn't know how to *get* to. Early "interactive fictions" and text-based computer games sometimes had this problem, nicknamed "guess-the-verb" (for instance, a 1978 game called *Adventureland* required the user to somehow know to type the ungrammatical command "unlight" to extinguish a lantern). It might be fair to say that therapy bots are to therapy books what interactive fictions are to novels.

vidually and idiosyncratically and painfully, has to spend some years working through it. The disease scales; the cure does not.

But is that always necessarily so? There are times when our bodies are sufficiently different from others' bodies that we have to be treated differently by doctors, though this doesn't frequently go beyond telling them our allergies and conditions. But our minds: How similar are they? How site-specific does their care need to be?

Richard Bandler is the co-founder of the controversial "Neuro-Linguistic Programming" school of psychotherapy and is himself a therapist who specializes in hypnosis. One of the fascinating and odd things about Bandler's approach—he's particularly interested in phobias—is that he *never finds out* what his patient is afraid of. Says Bandler, "If you believe that the important aspect of change is 'understanding the roots of the problem and the deep hidden inner meaning' and that you really have to deal with the content as an issue, then probably it will take you years to change people." He doesn't *want* to know, he says; it makes no difference and is just distracting. He's able to lead the patient through a particular method and, apparently, cure the phobia without ever learning anything about it.

It's an odd thing, this: we often think of therapy as *intimate*, a place to be understood, profoundly understood, perhaps better than we ever have been. And Bandler avoids that understanding like—well, like ELIZA.

"I think it's extremely useful for you to behave so that your clients come to have the illusion that you understand what they are saying verbally," he says. "I caution you against accepting the illusion for yourself."

Supplanted by Pure Technique

I had thought it essential, as a prerequisite to the
very possibility that one person might help another
learn to cope with his emotional problems, that the
helper himself participate in the other's experience of

those problems and, in large part by way of his own empathic recognition of them, himself come to understand them. There are undoubtedly many techniques to facilitate the therapist's imaginative projection into the patient's inner life. But that it was possible for even one practicing psychiatrist to advocate that this crucial component of the therapeutic process be entirely supplanted by pure technique—that I had not imagined! What must a psychiatrist who makes such a suggestion think he is doing while treating a patient, that he can view the simplest mechanical parody of a single interviewing technique as having captured anything of the essence of a human encounter?

—JOSEPH WEIZENBAUM

The term method itself is problematic because it suggests the notion of repetition and predictability—a method that anyone can apply. Method implies also mastery and closure, both of which are detrimental to invention.

—JOSUÉ HARARI AND DAVID BELL

Pure technique, Weizenbaum calls it. This is, to my mind, the crucial distinction. "Man vs. machine" or "wetware vs. hardware" or "carbon vs. silicon"–type rhetoric obscures what I think is the crucial distinction, which is between *method* and method's opposite: which I would define as "judgment," "discovery,"[3] "figuring out," and, an idea that we'll explore in greater detail in a couple pages, "site-specificity." We are replacing people not with *machines,* nor with *computers,* so much as with *method.* And whether it's humans or computers carry-

---

3. Glenn Murcutt, whom we'll hear more from later in this chapter: "We are taught that creativity is the most important thing in architecture. Well, I don't believe that. I think that the creative process leads to discovery, and discovery is the most important thing."

ing that method out feels secondary. (The earliest games of computer chess were played without computers. Alan Turing would play games of "paper chess" by calculating, by hand, with a pencil and pad, a move-selection algorithm he'd written. Programming this procedure into a computer merely makes the process go faster.) What we are fighting for, in the twenty-first century, is the continued existence of conclusions not already foregone—the continued relevance of judgment and discovery and figuring out, and the ability to continue to exercise them.

Reacting Locally

"Rockstar environments develop out of trust, autonomy, and responsibility," write programmers and business authors Jason Fried and David Heinemeier Hansson. "When everything constantly needs approval, you create a culture of nonthinkers."

Fellow business author Timothy Ferriss concurs. He refers to micromanagement as "empowerment failure," and cites an example from his own experience. He'd outsourced the customer service at his company to a group of outside representatives instead of handling it himself, but even so, he couldn't keep up with the volume of issues coming in. The reps kept asking him questions: Should we give this guy a refund? What do we do if a customer says such and such? There were too many different cases to make setting any kind of procedure in place feasible, and besides, Ferriss didn't have the *experience* necessary to decide what to do in every case. Meanwhile, questions kept pouring in faster than he could deal with them. All of a sudden he had an epiphany. You know who *did* have the experience and *could* deal with all these different unpredictable situations? The answer was shamefully obvious: "the outsourced reps themselves."

Instead of writing them a "manual," as he'd originally planned, he sent an email that said, simply, "Don't ask me for permission. Do what you think is right." The unbearable stream of emails from the reps to Ferriss dried up overnight; meanwhile, customer service at the company improved dramatically. "It's amazing," he says, "how someone's IQ seems to double as soon as you give them responsibility and indicate that you trust them." And, as far too many can attest, how it halves when you take that responsibility and trust away.

Here in America, our legal system treats corporations, by and large, as if they were people. As odd as this is, "corporation" is etymologically bodily, and bodily metaphors for human organizations abound just about everywhere. There's a great moment in the British series *The Office* when David Brent waxes to his superior about how he can't bring himself to make any layoffs because the company is "one big animal. The guys upstairs on the phones, they're like the mouth. The guys down here [in the warehouse], the hands." His boss, Jennifer, is senior management: the "brain," presumably. The punch line of the scene is, naturally, that David—who of course is the most layoff-worthy yet is the one in charge of the layoffs—can't figure out what organ *he* is, or what role *he* plays in the organization.

But there's a deeper point worth observing here too, which is that we create a caste system at our companies that mimics the caste system we create with respect to our own bodies and selves. My hands are *mine*, we say, but my brain is *me*. This fits in nicely with our sense of an inner homunculus pulling the levers and operating our body from a control room behind our eyeballs. It fits in nicely with Aristotle's notion that thinking is the most human thing we can do. And so we compensate accordingly.

I almost wonder if micromanagement comes from the same over-biasing of deliberate conscious awareness that led, both to and out of, the Turing machine model of computation underlying all of our computers today. Aware of everything, acting logically, from the top down, step-by-step. But bodies and brains are, of course, not like that at all.

Micromanagement and out-of-control executive compensation are odd in a way that dovetails precisely with what's odd about our rationalist, disembodied, brain-in-a-vat ideas about ourselves. When I fight off a disease bent on my cellular destruction, when I marvelously distribute energy and collect waste with astonishing alacrity even in my most seemingly fatigued moments, when I slip on ice and gyrate crazily but do not fall, when I unconsciously counter-steer my way into a sharp bicycle turn, taking advantage of physics I do not understand using a technique I am not even aware of using, when I somehow catch the dropped oranges before I know I've dropped them, when my wounds heal in my ignorance, I realize how much bigger I am than I think I am. And how much more important, nine times out of ten, those lower-level processes are to my overall well-being than the higher-level ones that tend to be the ones getting me bent out of shape or making me feel disappointed or proud.

Software development gurus Andy Hunt and Dave Thomas make the point that with a certain latitude of freedom and autonomy, a greater sense of *ownership* of a project emerges, as well as a sense of *artistry*; as they note, the stonemasons who helped build cathedrals were far from drones—they were "seriously high quality craftsmen."

> The idea of artistic freedom is important because it promotes quality. As an example, suppose you're carving a gargoyle up in the corner of this building. The original spec either says nothing or says you're making a straight on gargoyle just like these others. But you notice something because you're right there on the ground. You realize, "Oh look, if I curve the gargoyle's mouth this way, the rain would come down here and go there. That would be better." You're better able to react locally to conditions the designers probably didn't know about, didn't foresee, had no knowledge of. If you're in charge of that gargoyle, you can do something about that, and make a better overall end product.

> I tend to think about large projects and companies not as pyramidal/hierarchical, per se, so much as fractal. The level of decision making and artistry should be the same at every level of scale.

The corporate isn't always a great example of this. The corporeal is, as is another kind of organization with the body in its etymology—the U.S. Marine Corps. Consider this, from their classic handbook *Warfighting*:

> Subordinate commanders must make decisions on their own initiative, based on their understanding of their senior's intent, rather than passing information up the chain of command and waiting for the decision to be passed down. Further, a competent subordinate commander who is at the point of decision will naturally better appreciate the true situation than a senior commander some distance removed. Individual initiative and responsibility are of paramount importance.

In some sense this question of management style, of individual responsibility and agency, cuts not only across traditional divisions of "blue"- and "white"-collar work, but also between "skilled" and "unskilled" work. A formulaic mental process rigidly repeated time and again is not so different from a physical process repeated in this way. (That is to say, there's such a thing as thinking unthinkingly.) Likewise, a complex or sophisticated or learnèd process repeated time and again is not so different from a simple process repeated. More important than either of these distinctions, arguably, is the question of how much reacting locally or site-specificity, how much freshness in approach the job requires—or permits.

In March 2010, National Public Radio's *This American Life* did a segment on the joint General Motors and Toyota plant NUMMI. One of the biggest differences, it turned out, between the two companies was that at Toyota, "when a worker makes a suggestion that saves money, he gets a bonus of a few hundred dollars or so. Everyone's expected to be looking for ways to improve the production process. All the time. This is the Japanese concept of *kaizen,* continuous improvement." One American GM worker was part of a group that traveled to Japan to try building cars on the Toyota assembly line, and the difference in the experience stunned him:

> I can't remember any time in my working life where anybody asked for my ideas to solve the problem. And they literally want to know. And when I tell them, they listen. And then suddenly they disappear and somebody comes back with the tool that I just described. It's built, and they say, "Try this."

One of the results of this kind of participation is that "IQ-doubling" effect that Ferriss describes. You're not just *doing* something; you're doing that very human thing of simultaneously stepping back and considering the process itself. Another effect: *pride.* NUMMI's union leader, Bruce Lee, said he'd never felt about the cars he built like he did once he was participating in the process: "Oh, I was so proud of 'em you can't believe."

A Robot Will Be Doing Your Job

For the many, there is a hardly concealed discontent.
The blue-collar blues is no more bitterly sung than the
white-collar moan. "I'm a machine," says the spot-welder.
"I'm caged," says the bank teller, and echoes the hotel
clerk. "I'm a mule," says the steelworker. "A monkey
can do what I do," says the receptionist. "I'm less than
a farm implement," says the migrant worker. "I'm an
object," says the high-fashion model. Blue collar and
white call upon the identical phrase: "I'm a robot."

—STUDS TERKEL

The notion of computer therapists of course raises one of the major things that people think of when AI comes to mind: losing their jobs. Automation and mechanization have been reshaping the job market for several centuries at this point, and whether these changes have been positive or negative is a contentious issue. One side argues that machines take human jobs away; the other side argues that increased mechanization has resulted in economic efficiency that raises the standard of living for all, and that has released humans from a number of unpleasant tasks. The corollary to the "advance" of technology seems to be that familiar human "retreat," for better and for worse.

We call a present-day technophobe a "Luddite," which comes from a group of British workers who from 1811 to 1812 protested the mechanization of the textile industry by sabotaging mechanical looms:[4] this debate has been going on—in words and deeds—for centuries. But software, and particularly AI, changes this debate profoundly—because suddenly we see the mechanization of *mental* work. As Matthew Crawford argues in the 2009 book *Shop Class as Soulcraft*, "The new frontier of capitalism lies in doing to office work what was previously done to factory work: draining it of its cognitive elements."

I would also like to note something, though, about the *process* by which jobs once performed by humans get taken over by machines, namely, that there's a crucial intermediate phase to that process: where *humans* do the job *mechanically*.

Note that the "blue collar and white" workers complaining about their robotic work environments in Terkel's 1974 book *Working* are bemoaning not jobs they've *lost*, but the jobs they *have*.

This "draining" of the job to "robotic" behavior happens in many cases long before the technology to automate those jobs exists. Ergo, it must be due to capitalist rather than technological pressures. Once the jobs have been "mechanized" in this way, the much later process

---

4. In fact, the etymology of "sabotage," which comes from the French word *sabot*, meaning a type of wooden clog, is said (perhaps apocryphally) to be rooted in a story of workers destroying mechanical looms by throwing their shoes at, and into, them.

by which those jobs actually get taken over by machines (or, soon, AIs) seems like a perfectly sensible response, and, by that point, perhaps a relief. To my mind, the troubling and tragic part of the equation is the *first* half—the reduction of a "human" job to a "mechanical" one—and less so the second. So fears over AI would seem to miss the point.

Micromanagement; the *kaizen*-less assembly line; the over-standardization of brittle procedures and protocols . . . these problems are precisely the same problem, and pose precisely the same danger, as does AI. In all four cases, a robot will be doing your job. The only difference is that in the first three, the robot will be you.

"I Can't Communicate"

We're in the midst of an interesting moment for AI chatbots, which are finally starting to show commercial promise. Just recently, the Alaska Airlines website wanted me to chat with "Jenn" instead of using their customer service telephone number (I declined), and that's just the most recent example of many.[5] But before chatbots, there was, of course, the much-loathed automated telephone menu systems. And before that, human operators were behaving like chatbot automatons. Operator Heather Lamb, for instance, says in *Working*, "There are about seven or eight phrases that you use and that's it. 'Good morning, may I help you?' 'Operator, may I help you?' 'Good afternoon.'

---

5. See also, for instance, the Army chatbot SGT STAR, who gives very Army-like yet strangely endearing answers like the following: "Brian: What is Delta Force? SGT STAR: Delta Force is rumored to be the Army's counter-terrorist operations unit. In line with operational security we cannot discuss, nor ponder the unit's abilities." He even admonishes you in an Army-like (yet strangely endearing) way: "Brian: Wow, that sounds pretty badass. SGT STAR: OK Private, watch your language or I'll have to shut you down." The pathetic fallacy be damned—er, darned—I couldn't close the window without apologizing. "Sorry about that." "That's OK," he said, and I felt better.

'Good evening.' 'What number did you want?' 'Would you repeat that again?' 'I have a collect call for you from so-and-so, will you accept the charge?' 'It'll be a dollar twenty cents.' That's all you can say. A big thing is not to talk with a customer . . . I'm a communications person but I can't communicate."

I've called directory assistance a handful of times in recent years: the folks are as monosyllabic as possible, and brusque to the point of inhumanity. If my interaction with them is "human" in any way, it is only in the way that stepping on a stranger's toes on the bus and getting scowled at is "human." It's not their fault, of course—they're being forced to act like bots. In this particular case, I'll take the robot: at least I don't have to feel like a nuisance.

Now, if the people at directory assistance lived nearby, and could offer useful suggestions like "Oh, do you mean *Dave's* downtown or the *David's* up on Fifteenth?" or "But actually, if you're looking for a good steak my personal recommendation would be . . . ," that would be a different story entirely. But they don't live near where you're calling (scaling), they aren't given the time to engage with you (efficiency), and they can't really deviate from the script (pure technique).

Just today, I call to activate my new credit card and end up on the phone for a good ten minutes: the woman is getting snowed on in northern Colorado, wishing for milder weather, and I'm getting rained on in Seattle, wishing for a wintrier winter. Being from the Jersey shore, I've grown up accustomed to snowy winters and muggy summers. Sometimes I love the Northwest's temperateness; sometimes I miss the Northeast's intensity. The shore, she says, wow, I've never seen the ocean . . . And on from there. My roommate, passing through the living room, assumes it's an old friend on the line. Eventually the card is activated, and I snip the old one and wish her the best.

Maybe it's not until we experience machines that we appreciate the human. As film critic Pauline Kael put it, "Trash has given us an appetite for art." The inhuman has not only given us an *appetite* for the human; it's teaching us what it *is*.

Maggot Therapy

It's clear from all of this that AI is not really the enemy. In fact, it may be that AI is what extricates us from this process—and what identifies it. Friends of mine who work in software talk about how a component of their job often involves working directly on problems while simultaneously developing automated tools to work on those problems. Are they writing themselves out of a job? No, the consensus seems to be that they move on to progressively harder, subtler, and more complex problems, problems that demand more thought and judgment. They make their jobs, in other words, more human.

Likewise, friends of mine *not* in software—in PR, marketing, you name it—are increasingly saying to me: "Can you teach me how to program? The more you talk about scripting . . . I'm pretty sure I can automate half my job." In almost all cases they are right.

I think, quite earnestly, that all high school students should be taught how to program. It will give our next generation a well-deserved indignation at the repetitiveness and rule-bound-ness of some of the things they'll be asked to do. And it will also give them the solution.

You can almost think of the rise of AI not as an infection or cancer of the job market—the disease is *efficiency*—but as a kind of maggot therapy: it consumes only those portions that are no longer human, restoring us to health.

Art Cannot Be Scaled

Aretê . . . *implies a contempt for efficiency—or rather
a much higher idea of efficiency, an efficiency which
exists not in one department of life but in life itself.*

—H. D. F. KITTO, *THE GREEKS,* QUOTED
IN ROBERT PIRSIG'S *ZEN AND THE ART
OF MOTORCYCLE MAINTENANCE*

A farm equipment worker in Moline complains that
the careless worker who turns out more that is bad
is better regarded than the careful craftsman who
turns out less that is good. The first is an ally of the
Gross National Product. The other is a threat to
it, a kook—and the sooner he is penalized the bet-
ter. Why, in these circumstances, should a man work
with care? Pride does indeed precede the fall.

—STUDS TERKEL

French prose poet Francis Ponge's *Selected Poems* begins with the following: "Astonishing that I can forget, forget so easily and for so long every time, the only principle according to which interesting works can be written, and written well." Art cannot be scaled.

I remember my undergraduate thesis adviser, the fiction writer Brian Evenson, saying that writing books has never gotten any easier for him, because as he gets more adept at producing a certain kind of work, he grows, at an identical rate, less satisfied with repeating those old methods and practices that were so successful in the past. He won't let his old models scale. He won't let his job get any easier. To me that's exhilarating, the economy-in-all-senses-of-the-word-be-damned battle cry of the artist.

Ponge continues, "This is doubtlessly because I've never been able to define it clearly to myself in a conclusively representative or memorable way." Perhaps the issue is, indeed, that what makes for good art eludes description, will remain, by its very nature or by its relationship *to* description, forever ineffable. But this misses the more important point. I doubt Ponge would scale his production if he could. Evenson clearly won't. As former world chess champion Garry Kasparov puts it, "The minute I begin to feel something has become repetitive or easy, I know it's time to quickly find a new target for my energy." If, say, a musician like Carter Beauford is effortlessly and ceaselessly inventive on the drums, it is in part because he resolutely refuses to bore himself.

I remember having a conversation with composer Alvin Singleton, over dinner at an artists' colony last year. He told me about a clever title he'd used on one of his pieces, and I jokingly suggested a clever twist on the title, I think some kind of pun, be the title of his next piece. I expected a chuckle; instead, he was suddenly serious. "No, I would never use the same idea twice." For him this was no joke. Later I talked to him about how whenever I try to write music, the first thirty to forty-five seconds comes naturally to me, but then I get stuck. I wondered if, for him, whole songs just spring to mind the way short riffs and vamps spring to my mind. The answer was a steadfast negative. "What you call 'being stuck,' " he said, a twinkle in his eyes, "I call composing."

Site-Specificity

*I know when I'm working that the very first time I get
something right it's righter than it will ever be again.*

—TWYLA THARP

One of my friends, an architecture graduate student, was telling me about a famous architect—Australia's Pritzker Prize winner Glenn Murcutt—who's notoriously opposed to any kind of scaling-up of his output. The Pritzker jury clearly took note, saying, "In an age obsessed with celebrity, the glitz of our 'starchitects,' backed by large staffs and copious public relations support, dominates the headlines. As a total contrast, [Murcutt] works in a one-person office on the other side of the world . . . yet has a waiting list of clients, so intent is he to give each project his personal best." Murcutt himself doesn't find this scale-restraint, rare though it is, odd in the slightest. "Life is not about maximizing everything," he says. His wariness of scaling applies not only to his own operation but to the designs themselves. "One of the great problems of our period is that we've developed tools that allow rapidity, but rapidity and repetitiveness do not lead to right solutions. Perception gives us right solutions."

A fellow Pritzker-winner, French architect Jean Nouvel, concurs. "I think that one of the disasters in the urban situation today is what I call the generic architecture. It's all these buildings, parachuted into every city in the world now—and now with the computer you have a lot of facilities to do that. It's very easy. You can put three stories more or you can have a building be a little bit wider—but it's exactly the same building."

For Nouvel, too, the enemy is scaling (made nearly effortless by the computer), and the solution is found in perception. "I fight for specific architecture against generic architecture," he says. "I try to be a contextual architect . . . It's always, for a building, why this building has to be like *this*. What I can do here, I cannot—I cannot do in another place."

"It's great arrogance in an architect to think one can build any-where appropriately," says Murcutt. "Before starting any project I ask: What's the geology, what's the geomorphology, what's the history, where does the wind come from, where does the sun come from, what are the shadow patterns, what's the drainage system, what's the flora?"

Of course not *every* wheel has to be reinvented—Murcutt allegedly memorized catalogs of standard building components early in his career, stocking his imagination with available details, priming it to look for new ways to use them. It is paramount that those uses be site-specific.

"I think every site, every program, has a right to a specific work, to a complete involvement of the architect," Nouvel says. It's better for the work, better for the site—but, no less importantly, it's better for the *architects*. They, too, have a right to "complete involvement" in their work.

Most people, though, are not so involved—whether it is because they are prevented from doing so by the structure of their job or because (as with firms "parachuting" cloned buildings into city after city) they are complacent, thinking the problem already solved. For me, though, complacency—because it is a form of disengagement—is a whisker away from despair. I don't want life to be "solved"; I don't want it to be solv*able*. There is comfort in method: because we don't always have to reinvent everything at every minute, and because our lives are similar enough to others' lives, the present similar enough to the past that, for example, wisdom is possible. But wisdom that feels final rather than provisional, an ending rather than starting point, that doesn't ultimately defer to an even larger mystery is deadening. I won't have it. *Perception* gives us right solutions.

I think of site-specificity as a state of mind, a way of approaching the world with senses attuned. The reason to wake up in the morning is not the *similarity* between today and all other days, but the *difference*.

Site-Specificity in Conversation

If I speak in the tongues of men and of angels,
but have not love, I am only a resounding
gong or a clanging cymbal.

—1 CORINTHIANS 13:1

And if you're just operating by habit,
then you're not really living.

—*MY DINNER WITH ANDRE*

Many of my all-time favorite movies are almost entirely verbal. The entire plot of *My Dinner with Andre* is "Wallace Shawn and Andre Gregory eat dinner." The entire plot of *Before Sunrise* is "Ethan Hawke and Julie Delpy walk around Vienna." But the dialogue takes us everywhere, and as Roger Ebert notes, of *My Dinner with Andre*, these films may be paradoxically among the most *visually* stimulating in the history of the cinema:

One of the gifts of "My Dinner with Andre" is that we share so many of the experiences. Although most of the movie literally consists of two men talking, here's a strange thing: We do not spend the movie just passively listening to them talk. At first, director Louis Malle's sedate series of images (close-ups, two-shots, reaction shots) calls attention to itself, but as Gregory continues to talk, the very simplicity of the visual style renders it invisible. And like the listeners at the feet of a master storyteller, we find ourselves visualizing what Gregory describes, until this film is as filled with visual images as a radio play—more filled, perhaps, than a conventional feature film.

Sometimes there's so much that needs to be said that the literal "site" disappears, becomes, as in *My Dinner with Andre,* "invisible." Shawn and Gregory's restaurant seems to be one of those restaurants that's "nice" to the point of invisibility, as if any "holds" on their attention would be a distraction, as if (and this was Schopenhauer's view) happiness consisted merely in the eradication of all possible irritants and displeasures, as if the goal were to make the diners consent that they'd enjoyed themselves primarily through the impossibility of any particular criticism. The restaurant makes it a point to stay out of the way.[6] And in this particular movie and situation, it works brilliantly, because Shawn and Gregory just riff and riff, out to infinity. (And in fact, the next time they "notice" the restaurant, it ends their conversation—and the film.)

---

6. There's a very real downside to this particular style of luxury. Princeton psychologist Daniel Kahneman notes that arguments between couples are worse in luxury cars than in crappy cars, *precisely* for the things that they've paid top dollar for about the car. It's soundproof, so the noises of the world don't get in. It's comfortable, it runs smoothly and quietly, the suspension treats you gingerly. And so the argument goes on and on. Most disagreements are not 100 percent resolvable so much as they can be converted into more or less satisfactory compromises that then assume a lower priority than the other issues of life. They are terminated more from without than from within. Carl Jung puts it nicely: "Some higher or wider interest appeared on the person's horizon, and through this broadening of his or her outlook the unsolvable problem lost its urgency." Interruptions can be helpful.

But remember, Gregory and Shawn are old friends who haven't seen each other in years; Delpy and Hawke are just, like Turing test participants, starting from scratch. It's telling that in the sequel, *Before Sunset,* they walk around Paris, but Paris is much more "invisible" to them than Vienna was. They have, *themselves,* become the site.

Part of what makes language such a powerful vehicle for communicating "humanly" is that a good writer or speaker or conversationalist will tailor her words to the *specifics* of the situation: who the audience is, what the rhetorical situation happens to be, how much time there is, what kind of reaction she's getting as she speaks, and on and on. It's part of what makes the teaching of a specific conversational "method" relatively useless, and part of what makes the language of some salesmen, seducers, and politicians so half-human. George Orwell:

> When one watches some tired hack on the platform mechanically repeating the familiar phrases . . . one often has a curious feeling that one is not watching a live human being but some kind of dummy . . . And this is not altogether fanciful. A speaker who uses that kind of phraseology has gone some distance towards turning himself into a machine. The appropriate noises are coming out of his larynx, but his brain is not involved as it would be if he were choosing his words for himself. If the speech he is making is one that he is accustomed to make over and over again, he may be almost unconscious of what he is saying, as one is when one utters the responses in church.[7]

---

7. I do tend on the whole to think that words mean most when they're composed freshly. E.g., that the sinner's improvised confession in the booth does more, means more, than the umpteen un-site-specific Hail Marys he's prescribed and recites by rote.

This is also what makes the meeting of strangers such a difficult place to defend against machine imitation: we don't yet *have* the kind of contextual information about our audience that would enable us to speak more responsively and uniquely.

In these moments, where the metaphorical site-specificity of audience temporarily fails us, *literal* site-specificity may be able to help.

In *Before Sunrise,* set in Vienna when Delpy and Hawke are strangers and don't even know what to ask each other, the city itself spurs, prompts, and anchors the dialogue by which they come to know each other. Professional interviewers talk about how helpful these site-specifics can be. Their stance on the *My Dinner with Andre / Before Sunrise* question turns out to be surprisingly firm. "It's one thing to have lunch with a celebrity in some nice restaurant . . . It's another when you can just follow that person around for a while and see them in action," says *Rolling Stone*'s Will Dana in *The Art of the Interview,* where the *New York Times*'s Claudia Dreifus concurs. "Certainly no restaurants."

At a Turing test, one of the things that every bot author fears is that the judge will want to talk about the immediate environment. What color is Hugh Loebner's shirt? What do you think of the art in the lobby? Have you tried any of the food vendors outside? It's extremely hard to update your program's script that close to the contest itself.

I should think that a non-localized Turing test, in which participants don't gather at a particular city and building but are connected to other humans (and bots) *at random* around the world, would be much, much more difficult for the humans.

The conversational encounters of the Loebner Prize are frequently compared to those between strangers on a plane; I think part of the reason this analogy appeals so much to the contest's organizers (who of course are hoping for a close fight) is that planes are so *alike.* And seatmates have nothing really in common. But of course on a real plane the first thing you talk about is the city you're leaving and the city you're flying to. Or the book you notice in their lap. Or the funny way the captain just said such and such. Site-specificity manages to get its foot in the door.

When a round of the Loebner Prize in Brighton was delayed by fifteen minutes, I smiled. Any deviation from the generic plays into the humans' hands. As the round finally got under way and I began to type, the delay was the very first thing I mentioned.

The Problem of Getting the Part

Every child is born an artist. The trouble is how to stay one as you grow up.

—PABLO PICASSO

Most of the classes an actor takes are about how to get the part, and how to prepare for opening night. For the college actor, that's about all there *is*—the longest run of any show is typically two weekends, with many only scheduled for one or two shows total. The film actor is in a similar position: get it right and never do it again. But the professional stage actor might land a role that he or she has to reprise eight times a week for months, even years. How do you still feel like an artist in the tenth performance? The twenty-fifth? The hundredth?

(As Mike LeFevre in Studs Terkel's *Working* puts it: "It took him a long time to do this, this beautiful work of art. But what if he had to create this Sistine Chapel a thousand times a year? Don't you think that would even dull Michelangelo's mind?")

Art doesn't scale.

This problem fascinates me, in part because I believe it to be the problem of living.

How do you still feel creative when you're creating more and more of the same thing? Well, I think the answer is that you can't. Your only choice is to create more and more different things.

One of my favorite theatrical events, and a yearly ritual since I've lived on the West Coast, is Portland's Anonymous Theatre. One night only, one performance only: the director has cast the show and rehearsed with each of the cast members for weeks—that is, rehearsed with each of them *separately*. They don't know who else is in the show, and they won't meet until they meet onstage. There is no set blocking—they must react dynamically and without much of a preset plan. There's no rehearsal-dug ruts or habits between the actors—they must build their rapport and repartee in real time, with all eyes watching. It's fascinating and sublime to watch this happen.

When I talk to friends of mine who are actors, they say that this is more or less the answer that the actor in the long-running show must find, in his or her own way. How do you deviate? How do you make it a new show?

It'd be tempting to think that you spend a certain amount of time learning what to do, and the rest of the time knowing what you're doing, and simply doing it. The good actor will refuse to let this happen to him. The moment it does, he dies. A robot takes his place.

I think of neoteny, of my younger cousin, age four or so, careening into walls, falling over, getting up, and dashing off in the next direction. Of the fact that kids are much quicker studies at learning to ski because they're not afraid to fall. Fail and recover.

For the architect, it's site-specificity; for the actor and the musician, it's *night*-specificity. My friend Matt went to see a songwriter that he and I admire a lot, and I asked him what the show was like. Matt, unenthusiastic, shrugged: "He has a set list, and, you know, he plays it." It's hard to imagine what the artist *or* the audience gets out of that. A great counterexample would be a band like Dave Matthews Band, where one night a song is four minutes long and the next night, twenty. I admire the struggle implicit in that—and the risk. There must be this constant gravity toward finding what works and sticking with it, toward solidifying the song, but no, they *abandon* what works as righter than it will ever be again, or come back to it the next night, but only to depart and try something else that could well fail. This is how you stay an artist as you grow up. And both they and the fans get a once-in-a-lifetime thing.

I suppose when you get down to it, everything is always once in a lifetime. We might as well act like it.

I saw my first opera this past year: *La Traviata*, starring soprano Nuccia Focile in the lead role. The program featured an interview with her, and the interviewer writes, "It's those unexpected moments that blindside a singer emotionally, Focile feels. In performance, a different phrasing of a word can suddenly take the involved singer by surprise and make her gulp or blink away tears." Focile seems to think of these moments as hazards, saying, "I must use my technical base to approach certain phrases, because the emotion is so great that I get involved too much." As a professional, she wants to sing consistently. But as a human, the fine attention to and perception of the tiny uniquenesses from night to night, the cracks in technique where we *get* involved, get taken by surprise, gulp, feel things freshly again—these are the signs we're alive. And our means of staying so.

5. Getting Out of Book

*Success in distinguishing when a person is lying
and when a person is telling the truth is highest
when . . . the lie is being told for the first time;
the person has not told this type of lie before.*

—PAUL EKMAN

For Life is a kind of Chess . . .

—BENJAMIN FRANKLIN

How to Open

Entering the Brighton Centre, I found my way to the Loebner Prize competition. Stepping into the contest room, I saw rows of seating where a handful of audience members had already gathered, and up front what could only be the bot programmers worked hurriedly, plugging in tangles of wires and making the last flurries of keystrokes. Before I could get too good a look at them, or they at me, the test's organizer this year, Philip Jackson, greeted me and led me behind a velvet curtain to the confederate area. Out of view of the audience and the judges, four of us sat around a table, each at a laptop set up specifically for the test: Doug, a Canadian linguistics researcher for Nuance Communications; Dave, an American engineer working for Sandia National Laboratories; Olga, a speech-research graduate student from

South Africa; and myself. As we introduced ourselves, we could hear the judges and audience members slowly filing in, but couldn't see them around the curtain. A man flittered by in a Hawaiian shirt, talking a mile a minute and devouring finger sandwiches. Though I had never met him before, I knew instantly he could only be one person: Hugh Loebner. Everything was in place, we were told, between bites, and the first round of the test would be starting momentarily. We four confederates grew quiet, staring at the blinking cursors on our laptops. I tried to appear relaxed and friendly with Dave, Doug, and Olga, but they had come to England for the speech technology conference, and were just here this morning because it sounded interesting. I had come all this way just for the test. My hands poised hummingbird-like over the keyboard, like a nervous gunfighter's over his holsters.

The cursor, blinking. I, unblinking.

Then all at once, letters and words began to materialize—

`Hi how are you doing?`

The Turing test had begun.

And all of a sudden—it was the strangest thing. I had the distinct sensation of being *trapped*. Like that scene in so many movies and television shows where the one character, on the brink of death or whatever, says, breathlessly, "I have something to tell you." And the other character always, it seems to me, says, "Oh my God, I know, me too. Do you remember that time, when we were scuba diving and we saw that starfish that was curled up and looked like the outline of South America, and then later, when I was back on the boat and peeling my sunburn, I said that it reminded me of this song, but I couldn't remember the name of the song? It just came to me today—" And the whole time we're thinking, *Shut up, you fool!*

I learned from reading the Loebner Prize transcripts that there are two types of judges: the small-talkers and the interrogators. The latter are the ones that go straight in with word problems, spatial-reasoning questions, deliberate misspellings . . . They're laying down a verbal obstacle course and you have to run it. This type of thing is extraordinarily hard for programmers to prepare against, because anything goes—and this is (a) the reason that Turing had language, and conversation, in mind as his test, because it is really, in some sense, a test of everything, and (b) the kind of conversation Turing seemed to have envisioned, judging from the hypothetical conversation snippets in his 1950 paper. The downside to the give-'em-the-third-degree approach is that there's not much room to *express* yourself, personality-wise. Presumably, any attempts to respond idiosyncratically are treated as coy evasions for which you get some kind of Turing-test demerits.

The small-talk approach has the advantage that it's easier to get a sense of who a person *is*—if there indeed *is* a person, which is, of course, *the* if of the conversation. And that style of conversation comes more naturally to layperson judges. For one reason or another, it's been explicitly and implicitly, at various points in time, encouraged among Loebner Prize judges. It's come to be known as the "strangers on a plane" paradigm. The downside of this is that these types of conversations are, in some sense, uniform: familiar in a way that allows a programmer to anticipate a number of the questions.

So here was a small-talk, stranger-on-a-plane judge, it seemed. I had this odd sensation of being in that classic film/TV position. "I have something to tell you." But that something was . . . myself. The template conversation spread out before me: *Good, you? / Pretty good. Where are you from? / Seattle. How about yourself? / London. / Oh, so not such a far trip, then, huh? / Nope, just two hours on the train. How's Seattle this time of year? / Oh, it's nice, but you know, of course the days are getting shorter . . .* And more and more I realized that it, the conversational boilerplate, every bit as much as the bots, was the enemy. Because it—"cliché" coming from a French onomatopoeia for the printing process, words being reproduced without either alteration or understanding—is what bots are made of.

I started typing.

`hey there!`

Enter.

`i'm good, excited to actually be typing`

Enter.

`how are you?`

Enter.

Four minutes forty-three seconds. My fingers tapped and fluttered anxiously.

I could just feel the clock grinding away while we lingered over the pleasantries. I felt—and this is a lot to feel at "Hi, how are you doing?"—this desperate urge to get off the script, cut the crap, cut to the chase. Because I knew that the computers could do the small-talk thing; it'd be playing directly into their preparation. How, I was thinking as I typed back a similarly friendly and unassuming greeting, do I get that lapel-shaking, *shut-up-you-fool* moment to happen? Once those lapels were shaken, of course, I had no idea what to say next. But I'd cross that bridge when I got there. If I got there.

Getting Out of Book

The biggest AI showdown of the twentieth century happened at a chessboard: grandmaster and world champion Garry Kasparov vs. supercomputer Deep Blue. This was May 1997, the Equitable Building, thirty-fifth floor, Manhattan. The computer won.

Some people think Deep Blue's victory was a turning point for AI, while others claim it didn't prove a thing. The match and its ensuing controversy form one of the biggest landmarks in the uneasy and shifting relationship between artificial intelligence and our sense of self. They also form a key chapter in the process by which computers, in recent years, have altered high-level chess forever—so much so that in 2002 one of the greatest players of the twentieth century, Bobby Fischer, declared chess "a dead game."

It is around this same time that a reporter named Neil Strauss writes an article on a worldwide community of pickup artists, beginning a long process in which Strauss ultimately, *himself,* becomes one of the community's leaders and most outspoken members. Along the course of these experiences, detailed in his 2005 bestseller, *The Game,* Strauss is initially awed by his mentor Mystery's "algorithms of how to manipulate social situations." Over the course of the book, however, this amazement gradually turns to horror as an army of "social robots," following Mystery's method to a tee, descend on the nightlife of Los Angeles, rendering bar patter "dead" in the same ways—and for the same reasons—that Fischer declared computers to have "killed" chess.

At first glance it would seem, of course, that no two subjects could possibly be further apart than an underground society of pickup artists and supercomputer chess. What on earth do these two narratives have to do with each other—and what do they have to do with asserting myself as human in the Turing test?

The answer is surprising, and it hinges on what chess players call "getting out of book." We'll look at what that means in chess and in conversation, how to make it happen, and what the consequences are if you don't.

All the Beauty of Art

At one point in his career, the famous twentieth-century French artist Marcel Duchamp gave up art, in favor of something he felt was even more expressive, more powerful: something that "has all the beauty of art—and much more." It was chess. "I have come to the personal conclusion," Duchamp wrote, "that while all artists are not chess players, all chess players are artists."

The scientific community, by and large, seemed to agree with that sentiment. Douglas Hofstadter's 1980 Pulitzer Prize–winning *Gödel, Escher, Bach,* written at a time when computer chess was over twenty-five years old, advocates "the conclusion that profoundly insightful chess-playing draws intrinsically on central facets of the human condition." "All of these elusive abilities . . . lie so close to the core of human nature itself," Hofstadter says, that computers' "mere brute-force . . . [will] not be able to circumvent or shortcut that fact."

Indeed, *Gödel, Escher, Bach* places chess alongside things like music and poetry as one of the most uniquely and expressively human activities of life. Hofstadter argues, rather emphatically, that a world-champion chess program would need so much *"general* intelligence" that it wouldn't even be appropriate to call it a *chess* program at all. "I'm bored with chess. Let's talk about poetry," he imagines it responding to a request for a game. In other words, world-champion chess means passing the Turing test.

This was the esteem in which chess, "the game of kings," the mandatory part of a twelfth-century knight's training after "riding, swimming, archery, boxing, hawking, and verse writing," the game played by political and military thinkers from Napoleon, Franklin, and Jefferson to Patton and Schwarzkopf, was held, from its modern origins in fifteenth-century Europe up through the 1980s. Intimately bound to and inseparable from the human condition; expressive and subtle as art. But already by the 1990s, the tune was changing. Hofstadter: "The first time I . . . saw . . . a graph [of chess machine ratings over time] was in an article in *Scientific American* . . . and I vividly remember thinking to myself, when I looked at it, 'Uh-oh! The handwriting is on the wall!' And so it was."

A Defense of the Whole Human Race

Indeed, it wasn't long before IBM was ready to propose a meeting in 1996 between their Deep Blue machine and Garry Kasparov, the reigning world champion of chess, the highest-rated player of all time, and some say the greatest who ever lived.

Kasparov accepted: "To some extent, this match is a defense of the whole human race. Computers play such a huge role in society. They are everywhere. But there is a frontier that they must not cross. They must not cross into the area of human creativity."

Long story short: Kasparov stunned the nation by losing the very first game—while the IBM engineers toasted themselves over dinner, he had a kind of late-night existential crisis, walking the icy Philadelphia streets with one of his advisers and asking, "Frederic, what if this thing is invincible?" But he hit back, hard, winning three of the next five games and drawing the other two, to win the match with an entirely convincing 4–2 score. "The sanctity of human intelligence seemed to dodge a bullet," reported the *New York Times* at the match's end, although I think that might be a little overgenerous. The machine had drawn blood. It had proven itself formidable. But ultimately, to borrow an image from David Foster Wallace, it was "like watching an extremely large and powerful predator get torn to pieces by an even larger and more powerful predator."

IBM and Kasparov agreed to a rematch a year later in Manhattan, and in 1997 Kasparov sat down to another six-game series with a *new* version of the machine: faster—twice as fast, in fact—sharper, more complex. And this time, things didn't go quite so well. In fact, by the morning of the sixth, final game of the rematch, the score is tied, and Kasparov has the black pieces: it's the computer's "serve." And then, with the world watching, Kasparov plays what will be the *quickest loss of his entire career.* A machine defeats the world champion.

Kasparov, of course, immediately proposes a 1998 "best out of three" tiebreaker match for all the marbles—"I personally guarantee I will tear it in pieces"—but as soon as the dust settles and the press walks away, IBM quietly cuts the team's funding, reassigns the engineers, and begins to slowly take Deep Blue apart.

Doc, I'm a Corpse

When something happens that creates a cognitive dissonance, when two of our beliefs are shown to be incompatible, we're still left with the choice of which one to reject. In academic philosophy circles this has a famous joke:

> A guy comes in to the doctor's, says, "Doc, I'm a corpse. I'm dead."
> The doctor says, "Well, are corpses . . . *ticklish*?"
> "Course not, doc!"
> Then the doctor tickles the guy, who giggles and squirms away. "See?" says the doctor. "There you go."
> "Oh my God, you're right, doc!" the man exclaims. "Corpses *are* ticklish!"

There's always more than one way to revise our beliefs.

Retreat to the Keep

Chess is generally considered to require "thinking" for skillful play; a solution of this problem will force us either to admit the possibility of a mechanized thinking or to further restrict our concept of "thinking."

—CLAUDE SHANNON

So what happened after the Deep Blue match?

Most people were divided between two conclusions: (1) accept that the human race was done for, that intelligent machines had finally come to be and had ended our supremacy over all creation (which, as you can imagine, essentially no one was prepared to do), or (2) what most of the scientific community chose, which was essentially to throw chess, the game Goethe called "a touchstone of the intellect," under the bus. The *New York Times* interviewed the nation's most prominent thinkers on AI immediately after the match, and our familiar Douglas Hofstadter, seeming very much the tickled corpse, says, "My God, I used to think chess required thought. Now, I realize it doesn't."

Other academics seemed eager to kick chess when it was down. "From a purely mathematical point of view, chess is a trivial game," says philosopher and UC Berkeley professor John Searle. (There are ten thousand billion billion billion billion possible games of chess for every atom in the universe.) As the *New York Times* explained:

> In "Gödel, Escher, Bach" [Hofstadter] held chess-playing to be a creative endeavor with the unrestrained threshold of excellence that pertains to arts like musical composition or literature. Now, he says, the computer gains of the last decade have persuaded him that chess is not as lofty an intellectual endeavor as music and writing; they require a soul.
>
> "I think chess is cerebral and intellectual," he said, "but it doesn't have deep emotional qualities to it, mortality, resignation, joy, all the things that music deals with. I'd put poetry and literature up there, too. If music or literature were created at an artistic level by a computer, I would feel this is a terrible thing."

In *Gödel, Escher, Bach,* Hofstadter writes, "Once some mental function is programmed, people soon cease to consider it as an essential ingredient of 'real thinking.' " It's a great irony, then, that he was among the first to throw chess out of the boat.

If you had to imagine one human being completely unable to accept *either* of these conclusions—(a) humankind is doomed, or (b) chess is trivial—and you're imagining that this person's name is "Garry Kasparov," you're right. Whose main rhetorical tear after the match, as you can well imagine, was, *That didn't count.*

Garry Kasparov may have lost the final game, he says. But Deep Blue didn't win it.

Strangely enough, it's *this* argument that I'm the most interested in, and the one I want to talk about. What seems at first to mean simply that he made an uncharacteristic blunder (which he did) actually has a very deep and altogether different meaning behind it. Because I think he means it *literally*.

Well, if Deep Blue didn't win it, who—or what—did?

This is the question that starts to take us into the really weird and interesting territory.

How a Chess Program Is Built

To answer it, we need to get into some briefly technical stuff about how chess computers work;[1] hopefully I can demystify a few things without going into soul-crushing detail.

Almost all computer chess programs work essentially the same way. To make a chess program, you need three things: (1) a way to represent the board, (2) a way to generate legal moves, and (3) a way to pick the best move.

Computers can only do one thing: math. Fortunately for them, a shockingly high percentage of life can be translated into math. Music is represented by air pressure values over time, video is represented by red, blue, and green intensity values over time, and a chessboard is just a grid (in computer jargon: "array") of numbers, representing what piece, if any, is on that square.[2] Compared to encoding a song,

---

1. Here I'm using the words "program" and "computer" interchangeably. There's actually a profound mathematical reason for this, and it's Turing of all people who found it. It's known as "computational equivalence," or the "Church-Turing thesis."
2. How something like an array of numbers is represented in computer memory—because you still have to get down from base 10 to base 2 (binary), and from base 2 down to electricity and/or magnetism, etc.—is something I leave to the interested reader to look up in a computer science or computer engineering textbook.

or a film: piece of cake. As is often true in computer science, there are nifty tricks you can do, and clever corners you can cut, to save time and space—in some cases, astonishingly much—but those don't concern us here.

Once the computer has a chessboard it can understand in its own language (numbers), it figures out what the legal moves are from a given position. This is also simple, in fact, rather boringly straightforward, and involves a process like: "Check the first square. If empty, move on. If not empty, check what kind of piece it is. If a rook, see if it can move one square left. If yes, check to see if it can move another square left, and so on. If not, see if it can move one square right . . ." Again, there are some clever and ingenious ways to speed this up, and if you're trying to take down a world champion, they become important—for example, Deep Blue's creator, IBM electrical engineer Feng-hsiung Hsu, designed Deep Blue's thirty-six-thousand-transistor move generator *by hand*—but we are not touching *that* level of detail with a ten-foot pole. If shaving off microseconds doesn't matter to you, then anything that tells you the moves will do the trick.

Okay, so we can represent the board and we can figure out what moves are possible. Now we need an algorithm to help us decide what move to make. The idea is this:

1. How do I know what my best move is? Simple! The best move is the one that, after you make the best countermove, leaves me in the best shape.
2. Well, but how do I know what your best countermove is? Simple! It's the one that, after *my* best reply, leaves *you* in the best shape.

   (And how do we know what my best reply is? Simple! See step one!)

You're starting to get the sense that this is a rather circular definition. Or not circular, exactly, but what computer scientists call *recursive*. A function that calls itself. This particular function, which calls itself, you might say, in reverse—what move makes things best, given the move that makes things worst, given the move that makes things best, etc.—is called a minimization-maximization algorithm, or "minimax algorithm," and it crops up virtually everywhere in the theory and the AI of games.

Well, if you're writing a program for tic-tac-toe, for instance, this isn't a problem. Because the game only has nine possible first moves, eight possible second moves, seven possible third moves, and so on. So that's nine factorial: $9! = 362,880$. That may seem like a big number, but that's kid stuff to a computer. Deep Blue, and this was fifteen years ago, could look at 300,000,000 positions *per second*.[3]

The idea is that if your "search tree" goes all the way to the end, then the positions resolve into win, loss, and draw, the results filter back up, and then you move. The thing about chess, though, is that the search tree *doesn't bottom out*. Searching the whole thing (10^{90} years was Claude Shannon's famous estimate) would take *considerably* longer than the lifetime (a paltry 13.73×10^9 years) of the universe.

So you have to pull it up short. There are very sophisticated ways of doing this, but the easiest is just to specify a maximum search depth at which point you just have to call off the dogs. (Calling off the search in some lines before you call it off in others is called *pruning*.) So how do you evaluate the position if you can't look any further ahead and the game's not over? You use something called a *heuristic*, which—barring the ability to consider any further moves or countermoves—is a kind of static guesstimate of how good that position seems, looking at things like who has more pieces, whose king is safer, and things like that.[4]

---

3. In contrast, this is how many Kasparov could look at: 3.
4. In essence, Deep Blue v. Kasparov was a matter of the former's vastly superior (by a factor of roughly 100 million) search speed versus the latter's vastly superior pruning and heuristics—which moves are worth looking at, and how they bode—what you might call *intuition*.

That's it: represent the board, find moves and search through the replies, evaluate their outcomes with a heuristic, and use minimax to pick the best. The computer can then play chess.

The Book

There is, however, one other major add-on that top computer programs use, and this is what I want to talk about.

Computer programmers have a technique called "memoization," where the results of frequently called functions are simply stored and recalled—much like the way most math-savvy people will, when asked, respond that 12 squared is 144, or that 31 is prime, without actually crunching the numbers. Memoization is frequently a big time-saver in software, and it's used in chess software in a very particular way.

Now, every time Deep Blue starts a game, from that standard initial position of chess, it gets cranking on those 300 million positions a second, looks around for a while, and makes its choice. Because it's a computer, and unless it has randomness specifically programmed in, it's likely to be the *same* choice. Every time.

Doesn't that seem like a lot of effort? A waste of electricity, from the environmental perspective alone?

What if we just calculated it *once* and memoized it, that is, *wrote down* what we decided—and then we simply *always did that*?

Well—and what if we started doing this game after game, position after position?

And what if we were able to upload databases of hundreds of thousands of grandmaster games and write them down, too?

And what if we looked at every professional game ever played by Garry Kasparov and did some 300 million positions/second analysis *ahead of time* of the best responses to the positions likely to come up against him? What if we did, in fact, several *months* of analysis ahead of time? And, while we're at it, what if we employed a secret team of human grandmasters to help the process along?

*This is hardly "cheating" since that is the way
chess masters play.*

—CLAUDE SHANNON, "PROGRAMMING A
COMPUTER FOR PLAYING CHESS"

But I'm making it sound more sinister than it was. First of all, Kasparov knew it was happening. Second, it's what all professional chess players do before all professional chess games: all the top players have "seconds," slightly weaker professional players who prepare analysis—personalized for the opponent—before a match or tournament. This, in addition to the massive repertoire of openings and opening theory that all top players know. That's how the game is played. And this corpus of pre-played positions, untold thousands, if not millions, of them, this difference between discovery and memory, is called *the book*.

The Two Ends: Openings and Endings

All chess games begin from the exact same position. Because there are only so many moves you can make from that starting position, games will naturally take a while to differentiate themselves. Thus, a database of, say, a million games will have a million examples of a player making a move from that initial configuration; all other configurations will be a diminishing fraction of that. The more popular lines[5] maintain that "density" of data longer, sometimes beyond 25 moves, whereas the more unpopular or offbeat lines might peter out much more quickly. (The world's top computer program in recent years, Rybka, supposedly has certain lines in the Sicilian "booked" up to 40 moves, or longer than many *games*—for instance, only one game in the Kasparov–Deep Blue rematch went to move 50.)

At the other side: once enough pieces have been taken off the board, you begin to arrive at situations where the computer can

---

5. (move sequences)

simply preprocess and record every single possible configuration of those pieces. For example, the simplest endgame would perhaps be king and queen vs. king—three pieces on the board. That makes for, let's see, $64 \times 63 \times 62 = 249,984$ positions (minus some illegal ones, like when the kings are touching), and if you factor in the horizontal and (in this case) vertical symmetry of the board, you're down to at most 62,496. Very manageable. Once you start adding pieces, it gets progressively hairier, but all positions involving six or fewer pieces have already been "solved." This includes positions like some rook-and-knight versus two-knight endings, where, for instance, every move leads to a draw with perfect play except one—with which the strong side can, with inhumanly perfect and unintuitive play, force a checkmate in 262 moves.[6] That used to be the record, actually; but now programmers Marc Bourzutschky and Yakov Konoval have found a seven-piece endgame with a forced mate in 517.

Positions like this seem to me to be vaguely evil—there's absolutely nothing you can say about them to make them make sense. No way to answer the question, "Why is that the best move?" other than by simply pointing to the move tree and saying, "I don't know, but that's what it says." There is no explanation, no verbal translation, no intuition that can penetrate the position. "To grandmasters, it may turn out that the dismaying message of the latest computer analysis is that *concepts do not always work* in the endgame" (emphasis mine), the *New York Times* wrote in 1986, and quoted U.S. Chess Federation administrator, and grandmaster, Arthur Bisguier: "We're looking for something esthetic in chess—logic is esthetic. This bothers me philosophically."[7]

Maybe, as a person who is always, always theorizing, always, always verbalizing, this is what disturbs me, too: there is no such thing to

---

6. Again, most *games* are over in 30 to 40.
7. The endgame database that they're referring to is the work, in the 1980s, of Ken Thompson at the very same Bell Laboratories in Murray Hill, New Jersey, where Claude Shannon wrote the breakthrough paper on computer chess in 1950.

be done. Computers' lightning-fast but unintuitive exploration of the game tree is known as the "brute force" method to game AI. This is what the "brute" in "brute force" means to me; this is what's brute about it. No theory. No words.

Anyway, these tables are known as "endgame databases" or "endgame tables," or "tablebases" or "telebases," but we're fairly safe in calling them "books." The principle—look up a position and play the prescribed move—is the same.

So: there's an opening book, and an ending book.

The middle game—where the pieces have moved around enough so that the uniform starting position is a distant memory, but there's enough firepower on the board so that the endgame is still far off—is where games are most different, most unique.

"The whole strategy in solving a game is to shrink that middle part until it disappears, so your beginning game and your endgame connect," says Rutgers University computer scientist Michael Littman.

"Fortunately," says Kasparov, "the two ends—opening research and endgame databases—will never meet."

The Two Ends: Greetings and Closings

Letter writing is a great example of how "opening book" and "endgame book" occur in human relations. Every schoolchild learns the greetings and closings of a letter. They're so formalized, ritualized, that, well, computers can do them. If I end a paragraph in MS Word, and begin a new paragraph "Your," I immediately see a tiny yellow box with "Yours truly" in it. If I hit return, it auto-completes. If I type "To who," the "m it may concern" auto-completes. "Dear S" gives me "ir or Madam," "Cord," "ially," etc.

We're literally taught this "opening book" and "ending book" in schools. Then we go through life with our ears out—whether we know it or not—for subtle trends and indications of connotation, of context, of fashion. "What's up" originally felt awkward to me as a kid, imitative and unnatural, inauthentic—I couldn't say it, I found, without some kind of quotation marks—but it became as natural to me as "Hi." Then I watched, a few years later, the same process happen to my parents: their first few "What's up's" seeming like pitiful attempts to be "hip," and then increasingly I found I barely noticed. Abbreviations and truncations like "What up" and "Sup," which seemed poised to take over the hip-greeting spot among the cool kids of my middle school, never quite made it. When I started negotiating the tricky formal-yet-informal, subordinate-yet-collegial space of email correspondence with my professors in college and then graduate school, my instinct was to close with "Talk to you soon," but gradually I began to wonder if that didn't feel like a coded demand for promptness on their part, which could be read as impolite. I observed, imitated, and quickly warmed to the closing "Best," which then over some months started to feel curt; at some point I switched to "All the best," which is my old standby these days. Etiquette is a bit like fashion: you never quite stop wising up to it.

And, I should add, it's a bit like fashion in that you should be careful where you get your advice from: this afternoon I idly Googled "business letter closings," and the top hit is a list that includes "Adios" and "Ta ta." I don't think so.

When I started translating poetry, and began an email correspondence with a Venezuelan writer in Spanish—which I do not use conversationally very much, and certainly had never written emails to native speakers in before—I quickly learned and began mimicking the author's "Estimado amigo" opening and "Salud y poesía!" or "Recibe un abrazo fraterno" closings. I remember looking up websites that showed traditional Spanish greetings and closings, but couldn't trust them: you never know what sounds too stiff or too casual, too old-fashioned or too newfangled—not to mention the effects of all the national and regional deviations among the Spanish-speaking world. This is seriously tricky territory. I want to personalize my own openings and closings, but it's a delicate thing: without a broader sample of what gets used, I'm hopelessly out of my depth. So I repeat back the few greetings and closings I know.

Try starting or, even worse, ending a conversation with a non-stock phrase. It feels almost unmanageably awkward, abrupt. You can barely *think* of something non-stock to say; if you do think it up, you can barely bring yourself to say it. The ritual tugs hard at us.

It's pretty clear that if you want to get a flavor for a conversation by sampling, say, one or two sentences at random, you don't sample from the beginning or the end; you sample from the middle.

It's odd, in a way, how much etiquette and social ritual—which is not the same thing as formality, as, for instance, the long and elaborately choreographed handshakes you used to see in the 1980s and '90s go to show—threaten, and it is a threat of sorts, to lengthen those "books."

"Of course, the culture writes . . . first, and then we write . . . ," says playwright Charles Mee.

And when I write a letter, my culture gets the first word and, other than my name, the last.

I can express myself through my *choice* of openings/greetings, but, in some sense, the words aren't mine. It isn't me saying them.

Fortunately the two ends will never meet, says Kasparov. But I think we've all—haven't we?—had that experience, the conversation that plays itself *entirely* out, the conversation where the formalities of the greeting reach all the way to meet the formalities of the closing, the conversation that at some level, as Kasparov puts it, "doesn't even count" because it has probably been had, verbatim, before.

As it turns out, this is the conversation that the bots want to have in a Turing test. The conversation that the confederates are, in a fairly tangible way, fighting against (if keystrokes can be blows). The statistical, cultural, ritual regularities of human interaction are the weaknesses that these machines exploit.

In the Gaps

Grandmaster games are said to begin with a *novelty,* which is the first move of the game that exits the book. It could be the fifth, it could be the thirty-fifth. We think about a chess game as beginning with move one and ending with checkmate. But this is not the case. The game begins when it gets out of book, and it ends when it goes into book. Like electricity, it only sparks in the gaps.[8]

The opening book, in particular, is massive. The game may end before you get out, but it doesn't *begin* until you do. Said differently, you may not get out alive; on the other hand, you're not alive *until* you get out.

Who Sacked My Knight? The Metaphysics of Book

My point is this. What would prevent— Mike, maybe
you can answer this question. What would prevent
Deep Blue from seeing the e6 pawn and just taking it if
Garry leaves it there so that it can get close to redress-
ing the material imbalance? After all, this sacrifice
it played was not played on its own, on its own voli-
tion, it was programmed in. Maybe by now Deep
Blue is thinking, when the new moves started on the
board, Who sacked my knight? *(Audience laughter.)*

—GRANDMASTER MAURICE ASHLEY,
COMMENTATOR DURING GAME 6

Like many people in the competitive chess world, both Deep Blue's developers and Garry Kasparov subscribe to a kind of metaphysics of the book: the book isn't the person. Deep Blue's lead engineer, Feng-hsiung Hsu, has quotations about wanting to play "the World Champion, not his home preparation against our openings"; Kasparov says the same thing about the machine.

So the book is not the person—and the book is not the game:

---

8. See, e.g., the figure from Jonathan Schaeffer's landmark *Science* paper on computer checkers, of his program Chinook's live search-tree analysis looking quite literally like a lightning bolt between the opening and ending book.

"Today's game doesn't even count as a game because probably it has been published before elsewhere." An extremely strong statement: a game of chess that fails to get out of book is not a game of chess at all.

A "real" game or no, here it is, with some of the original live commentary: Deep Blue (white) v. Kasparov (black), 1997, Game 6.

1. e4 c6 2. d4 d5 3. Nc3

GRANDMASTER YASSER SEIRAWAN: He [Kasparov]'s going into what looks like a Caro-Kann.

3 . . . dxe4 4. Nxe4 Nd7 5. Ng5

SEIRAWAN: I think it very likely that we will see one of those openings that are analyzed out for fifteen or twenty moves, because it's going to be now very hard for Kasparov to avoid those lines. In these types of positions, you don't want to play anything original, because you could get into a lot of trouble early. I think that he's going to play one of the main lines and be satisfied with the resulting position.

5 . . . Ngf6 6. Bd3

GRANDMASTER MAURICE ASHLEY: Opening a line for his bishop. And again Deep Blue is clearly in its opening book because it is playing very quickly.

6 . . . e6

ASHLEY: Kasparov trying to get his bishop quickly into the action; we anticipate the bishop on f8 moving shortly.

7. N1f3

[Seirawan begins to play the book response, 7 . . . Bd6, on the diagram board—]

7 . . . h6

ASHLEY: Instead of bringing out his bishop with Bd6, Kasparov has instead—

8. Nxe6

ASHLEY: Capturing on e6 instantly and Kasparov shook his head for a moment—

8 . . . Qe7 9. 0-0 fxe6 10. Bg6+ Kd8

ASHLEY: Kasparov is shaking his head as if something
disastrous has happened, his king being chased around
the board. Is it possible that Kasparov has played incorrect
theoretically?

SEIRAWAN: Yes, he has. He blundered. What he did is
he transposed moves. What I mean by that is this position
is quite well known, and you had witnessed me playing
the move Bf8-d6. The idea being that after Bd6, it's
standard for white to then play Qe2, and then after h6,
this sacrifice Nxe6 doesn't work because black has the
move Kf8 later.

ASHLEY: You mean after Nxe6?

SEIRAWAN: Capturing the knight, there's the check, the king
can go to f8. But playing h6 one move earlier, the sacrifice
that we've now seen, Nxe6, is possible. As far as I recall,
there was a famous game between [Julio] Granda Zúñiga,
Grandmaster from Peru, versus our very own Patrick Wolff.
And it was a very difficult game for black to play and it
became recognized that the move h6 was wrong.[9] And
Garry, as you saw his reactions, the moment that Deep
Blue played Nxe6 so very quickly and reached the position
they now have on the board, he was in just terror, distress.
Because he's—he recognizes that he's fallen for a well-
known opening trap.

ASHLEY: Is this over? Is it that simple? . . . How is this
possible, Yaz? . . .

---

9. Granda is known for being in all likelihood the strongest player, and possibly the only grandmaster, not to study opening theory. Against weaker players, his unpredictability gives him an edge, but against the world's elite any
inaccuracy or imprecision whatsoever in the opening will usually be fatal.
The irony is that in the game Seirawan is citing, the one that established "that
the move h6 was wrong," Granda, after tenaciously clinging on for dear life,
actually goes on to win.

SEIRAWAN: . . . One of the things that—and in fact I find most upsetting about this particular position is, if Garry Kasparov were to lose today's game, it's entirely conceivable this whole sacrifice and so on is just in Deep Blue's library, opening library, and it's done nothing—it may turn out it won't even have to play an original move if Garry chooses one of the variations that it has been programmed as a win for itself. Which would be very unfair, not only to the Deep Blue team and its research, but to Garry Kasparov as well.

And this is why Game 6 didn't count. Kasparov bungled his seventh move (7 . . . h6, intending 8 . . . Bd6, instead of the correct 7 . . . Bd6 first, followed by 8 . . . h6), falling into a well-known book trap. The machine looked up the position and delivered the withering knight sacrifice (8.Nxe6) straight from its tables. Kasparov eventually wrenched himself out of book with a novel, if desperate, defense (11 . . . b5), but it was too late—by the time Deep Blue, that is, the search, analysis, and move selection procedure, finally stepped in, it was only to deliver the coup de grâce.

I agree with him that Game 6 "didn't count." He could have defended better, he could have held on longer, but at heart Kasparov lost that game *in book*.[10] (One commenter online described it with the wonderful phrase "game six's all-book crusher.") Tripping and falling into a well on your way to the field of battle is not the same thing as dying on it.

And—and here's a more metaphysical assertion, the one that takes us back to the Turing test and back to ourselves—whoever or what-

---

10. Hsu claims in *Behind Deep Blue* that most chess programs of the day were programmed specifically to *avoid* 8.Nxe6, because, although the best move and only clear refutation of 7 . . . h6, it led to tricky follow-throughs. He argues that Deep Blue merely called Kasparov's bluff: that his 7 . . . h6 was played on the assumption that Deep Blue was muzzled in that line. "A $300,000 gamble," Hsu calls it. I see the logic, but I don't buy it. It's pretty clear Kasparov simply screwed up.

ever achieved that winning position over Kasparov *was not Deep Blue.* "Hey! Who sacked my knight!?" Ashley jokingly imagines Deep Blue wondering, as its analysis function finally kicks in. Indeed.

Deep Blue is only itself out of book; prior to that it is nothing. Just the ghosts of the game itself.

And so it is with us, I find myself saying.

The End of Memory's Rope

Some of my friends growing up were competitive players of the game 24 in middle school. The idea of the game was that there were these cards of four numbers—say, 5, 5, 4, 1—and you had to figure out a way to add, subtract, multiply, or divide them in order to produce the number 24: in this case, $5 \times (5 - 1) + 4$ would do it, for instance. New Jersey had an annual 24 tournament for the state's best middle schoolers, and they got to go. Well, one of the kids there had spent basically the entire month before the finals simply *memorizing the cards.* He announced this to the other competitors at his table with a smirk. The finals were timed—first person to shout out the correct answer gets the point—and so the insinuation was that no one would have a chance: they'd be *calculating,* and he'd be *remembering.* When the official casually announced, in her opening remarks, that new cards had been prepared specially for this event, my friends watched, and may have suppressed a smirk or two themselves, as all the color drained from the kid's face. The round began, and he got slaughtered.

Garry Kasparov observes that this memorization-based approach is alarmingly common among new players:

> Players, even club amateurs, dedicate hours to studying and memorizing the lines of their preferred openings. This knowledge is invaluable, but it can also be a trap . . . Rote memorization, however prodigious, is useless without understanding. At some point, he'll reach the end of his memory's rope and be

without a premade fix in a position he doesn't really understand . . .

In June 2005 in New York I gave a special training session to a group of the leading young players in the United States. I had asked them each to bring two of their games for us to review, one win and one loss. A talented twelve-year-old raced through the opening moves of his loss, eager to get to the point where he thought he'd gone wrong. I stopped him and asked why he had played a certain pawn push in the sharp opening variation. His answer didn't surprise me: "That's what Vallejo played!" Of course I also knew that the Spanish Grandmaster had employed this move in a recent game, but I also knew that if this youngster didn't understand the motive behind the move, he was already headed for trouble.

In *The Game,* Neil Strauss recounts trying to initiate a threesome with two women he'd recently befriended. Otherwise clueless, all he knows is what Mystery has told him he's done in the past—run a bath and ask for help washing your back. So Strauss goes through the same moves, with predictably disastrous results:

Now what?

I thought sex was supposed to automatically happen afterward. But she was just kneeling there, doing nothing. Mystery hadn't told me what I was supposed to do after asking them to wash my back. He'd just said take it from there, so I assumed the whole sex thing would unfold organically. He hadn't told me how to transition . . . And I had no idea. The last woman to wash my back was my mother, and that was when I was small enough to fit in the sink.

Strauss finds himself in an awfully strange situation, of course, because he's there in the tub . . . but he didn't actually *want* a bath—so it's the same "Who sacked my knight?" problem: "Who asked for the back wash?" When the authentic Neil "kicks in," how does he deal with the odd things he said and did when he was possessed by the book?

Death of the Game

Neil Strauss argues that the nightlife of L.A.'s Sunset Boulevard has been blighted by a new generation of pickup artists—which he refers to as "social robots"—who have taken the art out of flirtation by replacing genuine conversational ability with mere and breathtakingly elaborate "opening line" repertoire.[11] "International dating coach" Vin DiCarlo is compiling a database of text messages, thousands upon thousands, and cataloging their success ("even down to the punctuation") at prompting replies and dates. Increasingly-popular dating websites and books offer pre-scripted conversational openings, emphasizing memorization and repetition: "Once you have performed a particular story or routine dozens of times, you don't even have to think about what you are saying. Your mind is free for other tasks, such as planning the next move. You have already fully explored all the conversational threads that could possibly arise from this piece of material. It's almost like seeing the future."

The fascinating thing is that we now have a *measure* for the failure of these conversational approaches. That measure is the Turing test—because its programmers are, ironically, using many of the self-same approaches.

Not all conversational autopilot happens in courtship, though, and sometimes it overtakes those even with the best of intentions. News anchor and interviewer Ted Koppel bemoans, "It's amazing how many

---

11. The most famous of which is arguably the "jealous girlfriend opener," popular enough that by the end of *The Game* two women say to Strauss when he approaches them, "Let me guess. You have a friend whose girlfriend is jealous because he still talks to his ex from college. Like, every guy keeps asking us that. What's the deal?" See also, e.g., the "cologne opener," "Elvis opener," "who lies more opener," "dental floss opener" . . .

people come into an interview having already decided what their questions are going to be, having decided on the order of those questions, and then pay absolutely no attention to what the interviewee is saying. Frequently people reveal something about themselves in an interview, but if you don't follow up on it, it will be lost." Nor does this kind of unresponsive script-reciting and rule-following happen only to those who use it as a deliberate or half-deliberate strategy. I think we've all, at some moments or others, found ourselves running through the standard conversational patterns and "book" responses without realizing it, or have actively searched for ways to get a conversation out of book, but didn't know how.

The history of AI provides us not only with a *metaphor* for this process but also with an actual *explanation,* even a set of benchmarks—and, better than that, it also suggests a solution.

Our first glimpse at what that solution might look like comes from the world of checkers, one of the first complex systems to be rendered "dead" by its book—and this was the better part of a century before the computer.

Putting Life Back into the Game

Checkers hit rock bottom in Glasgow, Scotland, in 1863.

Twenty-one games of the forty-game world-championship match then being played between James Wyllie and Robert Martins were *the exact same game* from start to finish. The other nineteen games, too, began with the same opening sequence, dubbed the "Glasgow opening," and all forty were draws.

For checkers fans and organizers alike (you can only imagine the mind-blowingly insipid headlines this match must have generated, and how rankled the sponsors must have felt), the Wyllie-Martins 1863 match was the last straw. Opening theory, combined with top players' take-no-risks attitude, had ground top-level checkers to a halt.

But what was to be done? How to save a dying game, rendered static by the accumulation and calcification of collective wisdom?

You couldn't just *force* world-class checkers players to systematically not play the moves established as the correct ones—or could you?

Perhaps, if you didn't like how players were opening their games, you could simply *open their games for them.* This is exactly what checkers' ruling body began to do.

Starting in America around 1900, serious tournaments began operating with what's called the "two-move restriction." Before a match, the first two opening moves are chosen at random, and the players play two games from the resulting position, one from either side. This leads to more dynamic play, less reliance on the book, and—thank God—fewer draws. But after another generation of play, even the two-move restriction, with its 43 starting positions,[12] began to seem insufficient, and in 1934 it was upped to a *three*-move restriction with 156 different starting positions. Meanwhile, in an odd twist, classic checkers, with no move restrictions, has become itself a kind of *variant,* called "Go-As-You-Please."[13]

A new method of opening randomization, called "11-man ballot," where one of the twelve pieces is removed at random from either side, and *then* the two-move restriction is applied, is now starting to gain traction. The number of starting positions in 11-man ballot checkers is in the thousands, and while the three-move restriction remains, since 1934, de rigueur at the top levels of play, it seems likely that the future of serious checkers, such as there is one, lies there.

Even when organizers aren't forcing players to randomize their openings, there may be good strategic reasons to do so: one makes an admittedly, if only slightly, weaker move than the move opening theory prescribes, hoping to come out ahead by catching one's opponent off guard. Garry Kasparov popularized this notion, dubbed

---

12. A small handful of openings are excluded for being simply too bad for one of the players: in general, slightly off-balance openings are fine so long as each player gets a turn with the stronger side.

13. It's worth noting that out of the 156 legal starting configurations of the three-move restriction, top checkers program Chinook has only "solved" 34 of them. Go-As-You-Please, though, it has completely locked up.

"anti-computer chess," in his games against Deep Blue: "I decided to opt for unusual openings, which IBM would not have prepared for, hoping to compensate for my inferior position with superior intuition. I tried to present myself to the computer as a 'random player' with extremely odd playing characteristics."[14]

When computers play each *other*, the influence of the opening book is so dramatic and, frequently, decisive that the chess community began to distrust these games' results. If you wanted to buy commercial chess software to analyze your own games and help you improve, how could you know which chess programs were stronger? The ones with the deepest books would dominate in computer tournaments, but wouldn't necessarily be the best analytically. One answer, of course, would be simply to "unplug" the opening book from the analysis algorithm and have the two programs play each other while calculating from move one. But this, too, produces skewed results—picking good opening moves is a different beast from picking good middle- or end-game moves, and it'd be unfair, and irrelevant, to have programmers, for the purposes of winning these computer contests, spend weeks honing opening-move analysis algorithms when in practice (i.e., when the software has access to the opening book) this type of analysis will never be used.

It was English grandmaster John Nunn who first addressed this problem in the late 1990s by creating "test suites" of a half dozen or so unusual (that is, out of book), complex, and balanced middle-game positions, and having programs take turns playing from either side of each of these positions, for a total of a dozen games. The programs

---

14. Indeed, he virtually flabbergasted the commentators, opening the third game of the rematch with 1.d3, a move that is almost unheard of at the grandmaster level (over 43 percent of grandmaster games begin with 1.e4, the most popular; only one in five thousand starts with 1.d3). Jaws dropped. International master Mike Valvo: "Oh my God." Grandmaster Maurice Ashley: "A cagey move, a shock of shocks in this match. This match has everything." Grandmaster Yasser Seirawan: "I think we have a new opening move."

simply begin playing "in medias res"—lopping off the opening phase of the game altogether.

In the beginning of the twenty-first century, former world champion Bobby Fischer shared these concerns, horrified at the generations of new players using computers to help them memorize thousands of book openings and managing to get the better of players with genuine analytical talent.[15] Chess had become too much about opening theory, he said, too much "memorization and prearrangement." "The time when both players actually start thinking," he said, "is being pushed further and further in." He came to an even more dramatic conclusion than Kasparov and Nunn, however, concluding, "Chess is completely dead."

His solution, though, was quite simple: scramble the order of the pieces in the starting position. Given a few basic guidelines and constraints (to maintain opposite-colored bishops and the ability to castle), you're left with 960 different starting positions: enough to water down the opening book to near irrelevance. This version of the game

---

15. From a 2006 radio interview: "It's . . . degenerated down to memorization and prearrangement . . . Chess, you know, so much depends on opening theory. Champions of, say, the last century, the century before last, they didn't know nearly as much as, say, I do, and other players know, about opening theory. So, if you just brought them back, you know, from the dead, and they played cold, they wouldn't do well, because they'd get bad openings . . . Memorization is enormously powerful . . . Some kid of fourteen, today, or even younger, could get the opening advantage against [1921–27 world champion José Raúl] Capablanca or especially against the players of the previous century . . . And maybe they'd still be able to outplay the young kid of today, but maybe not . . . So it's really deadly. It's very deadly. That's why I don't like chess anymore . . . And, you know, [Capablanca] wanted to change the rules already, back in, I think, the '20s; he said chess was getting played out. And he was right. (Interviewer: 'It's even more so now.') Oh, now, now it's completely dead. It's a joke. It's all just memorization and prearrangement. It's a terrible game now. ('And the computers . . . ') Yeah. It's a very *un*-creative game. ('And everything is known, and there is nothing new.') Well . . . let's not exaggerate. But it . . . it's really dead."

goes by "Fischer Random," "Chess960," or just "960." Chess960 now has its own world championship, and many of the top world players of traditional chess also now play 960.

Proust Questionnaire as Nunn Test Suite

What of these rejuvenation efforts in checkers and chess? What might be the conversational equivalent? One is simply to be aware of what utterances occur frequently and attempt to step out of their shadow; for instance, "What's your favorite color?" appears on over two million web pages, says Google, but "What's your favorite brunch food?" appears on a mere four thousand. Correspondingly, Cleverbot has a very good answer for the former ("My favorite color is green"), but, lacking a specific brunch-related answer in its conversational book, must fall back on the more general "My favorite food is sea food." It's not a dead giveaway in a Turing test, but it's a start, and more than enough to raise suspicion.

For the first few years, the Loebner Prize competition used specific topics of discussion, one for each program and confederate—things like Shakespeare, the differences between men and women, and the Boston Red Sox. The idea was to reduce the "domain" of discourse to a limited range, and give the computers a handicap while bot technology was still fledgling. Ironically, while removing the topic restriction, as they did in 1995, makes the computers' task almost infinitely harder in *theory*, a glance at Loebner Prize transcripts suggests that it may have actually made their job *easier* in practice. Instead of gearing their software toward a specific topic area that changes each year, programmers spend year after year after year honing their software to start in the same way each time—with a friendly greeting and some small talk. I suspect the most difficult Turing test would be something akin to the 11-man ballot of checkers, or Chess960: a randomly assigned topic, chosen by the organizers just before the conversation begins. The computers that have regularly gotten the better of judges in recent years wouldn't stand the whisper of a chance.

And what of life? Is this kind of "opening randomization," out-of-book practice useful to us humans, already certain of each other's humanity but looking, nonetheless, to get in contact with new vectors of it? The more I think about it, the more I think it is.

An old friend and I, as teenagers, used to be in the habit of doing this on long car trips: confident of our ability to riff on anything and in any direction, we'd name a subject out of thin air—"corn" was one of them that I remember—and just figure out what we wanted to say about it. Likewise, I remember a childhood game I had with my father—again, usually in the car—where I'd give him a topic and he'd just start some kind of improvised yarn. One day it might be the Civil War ("You know, the Civil War wasn't actually very civil at all: there was spitting, name-calling . . ."), another day, flatbed trucks ("You know, flatbed trucks didn't always used to be flat: things were constantly rolling off, and no one knew what to do . . .").

At a friend's wedding recently, the bride and groom gave each guest a "Proust questionnaire" to fill out at some point during the evening. It was adapted from a list of questions common in nineteenth-century diary books and famously answered (twice) by writer Marcel Proust, first in 1896 as a young teen, and again at age twenty. (Various contemporary celebrities now answer it on the back page of *Vanity Fair* each month.) The questions include offbeat and revealing things like "On what occasion do you lie?" "What trait do you most deplore in yourself?" "When and where were you happiest?" and "How would you like to die?" My girlfriend and I filled them out, and then we traded forms and read each other's answers. We are both, I would venture to say, very open and forthright and forthcoming people, and fluent conversationalists—by which I mean to say, whatever emotional depths we hadn't yet plumbed in our relationship were simply a factor of time and/or of not always knowing the best verbal routes to get there. Reading that questionnaire was a stunning experience: the feeling, one of doubling in an instant our understanding of the other. Proust had helped us do in ten minutes what we'd taken ten months to do on our own.

Sparks

We don't always *have* to initiate a conversation with some offbeat never-before-heard utterance, of course. Grandmaster Yasser Seirawan, commentator for the Kasparov–Deep Blue match, in fact criticized Kasparov's decision to play strange openings:

> Well, the mythology of how to play against computers is, they're loaded to the gills with this fantastic database . . . and what we ought to do is immediately get them out of their opening library— . . . I think it's quite okay to play main-line openings.[16] Because why are the opening library's moves being played and why are they put into the computer? Well, the reason they are is because guys like Garry Kasparov are playing these fantastic moves that become established as the best opening moves. But Garry is constantly reinventing the opening book, so my attitude, if I were Garry, is to say "Look, I'm going to play main-line stuff, the stuff that the computer will play. I'll go right down—right down the primrose path, and I'll ambush the computer with an opening novelty that it's never seen." And he's not doing that. Instead he's saying, "I want a completely unique, original game as early as I possibly can."

The same can be said conversationally: the reason things get established as the "main lines" is that, by and large, *they work*. This isn't always true; for instance, Robert Pirsig gives a pretty trenchant takedown of "What's new?" in *Zen and the Art of Motorcycle Maintenance*, and Yaacov Deyo, the inventor of speed dating, had to go so far as to *ban* the question "So, what do you do for a living?" because it was so ubiquitous and so unproductive. Notice, though, that Seira-

---

16. I.e., the most popular, well-trodden, and well-studied ones, the ones with the largest and deepest "books."

wan's defense of "main-line" openings is contingent on the fact that there *will* be a deviation eventually.

Of course, in the five-minute Turing test (unlike a seven-hour chess match at world-championship time controls) we don't *have* an "eventually." If we tread the primrose path in a Turing test, we do it at our peril. Far better, I think, to bushwhack.

Fischer wanted the same thing from chess that Kasparov wanted in his match against Deep Blue, and the same thing that Strauss wants in bar flirtation. It's what *we* want, chatting with old friends, when our familiar opening book of "Hi!" "Hi! How are you?" "Good, how are you?" "Good!"—which is not so much a conversation per se as a means for *arriving* at one—gives pleasantly way to the expectedly unexpected, awaitedly idiosyncratic veers; it's what *anyone* wants from *any* conversation, and what artists want from their art: a way to breeze past formalities and received gestures, out of book, and into the real thing.

And the book, for me, becomes a metaphor for the whole of life. Like most conversations and most chess games, we all start off the same and we all end up the same, with a brief moment of difference in between. Fertilization to fertilizer. Ashes to ashes. And we spark across the gap.

6. The Anti-Expert

Existence and Essence; Human vs. Hole-Puncher

It's hard to say we're lucky when we face a crisis, but we at least have the luxury of knowing that action is called for—of being forced to move. The truest tests of skill and intuition come when everything looks quiet and we aren't sure what to do, or if we should do anything at all.

—GARRY KASPAROV

One of the classic thought experiments in existentialism is the difference between humans and hole-punchers, in other words, the difference between people and machines.

Here's the crucial thing. The *idea* of the hole-puncher exists before the hole-puncher exists. Before the hole-puncher you got at Staples, there was a hole-puncher factory built to make that hole-puncher to particular design specifications that someone drafted up and had in mind. Before the hole-puncher was the idea of paper, and holes, and of punching holes in paper, and of making a machine for that purpose. As soon as the machine exists, it is playing the part assigned it by its designers. You buy it and put paper in it, and it punches holes into your paper. This is its essence, and to use it as a doorstop or paperweight or hammer or cudgel is to go against the grain of this essence.

The essence of the hole-puncher precedes its existence. We humans are not like this, argue the existentialists. With us, existence comes first.

A human being, writes Jean-Paul Sartre, "exists, turns up, appears on the scene, and, only afterwards, defines himself." What defines us is that we *don't* know what to do and there *aren't* any revelations out there for us waiting to be found. Profoundly disoriented and lacking any real mooring, we must make it all up from scratch ourselves, each one of us, individually.[1] We arrive in a bright room, wet, bloody, bewildered, some stranger smacking us and cutting what had been, up to that point, our only source of oxygen and food. We have no idea what is going on. We don't know what we're supposed to do, where we're supposed to go, who we are, where we are, or what in the world, after all this trauma, comes next. We wail.

Existence without essence is very stressful. These are not problems that the hole-puncher can understand.

Existence-and-essence arguments like these are actually rather familiar to most early-twenty-first-century Americans, because they are basically the "intelligent design" debate going on in our school system. A human being *is* a designed thing, the intelligent-design camp says, and in that sense very *much* like a paper cutter or (their preferred metaphor) a pocket watch. Along with this comes the idea of *discovering* your own "design"/function/purpose as you go through life. In children's book form, it'd be the watch who, one day, learns that he was made to tell people the time. Our everyday idioms are full of such references: "Man, that guy was *born* to speed skate," we say of a trunk-thighed Olympian.

The existentialists would protest: purposes aren't discovered or

---

1. But what if that *is* humans' purpose? That process of definition, the very process of *finding* a purpose? Vonnegut writes, "Tiger got to hunt, bird got to fly / Man got to sit and wonder, 'Why, why, why?' " This would make the existentialists feel good, the way Aristotle's decision that contemplation is the highest activity of man made Aristotle feel good, but in this case it would undermine their argument.

found, because they don't exist ahead of us. Purpose, in their view, can never be found, but must be *invented*.

Of course, thighs *are* made to contract and move the legs. One of the intriguing things about the existentialist argument is that it is a kind of more-than-the-sum-of-the-parts argument. My bicep has a function. My cells' tRNA has a function. I don't.

(Interestingly, even the opponents of intelligent design, supporters of Darwinism, are sometimes guilty of making life sound more teleological, goal-oriented, than it is. Harvard zoologist Stephen Jay Gould, for instance, takes pains in his 1996 book *Full House* to show that it's inappropriate to point, as many do, to the emergence of a complex species like ourselves from a world that was mostly bacterial as evidence that there's any notion of biological "progress" at work in the world.[2])

But to start acknowledging the functions and capacities out of which we're built—organs, of course, have purposes—is to start to acknowledge the limits of that existentialist equation, and of our "total" freedom and ability to make choices and fashion our own existences. Existentialism is, in this way, classist. You don't worry about what to wear if you only have one outfit; you don't worry about what to do with your life if you only have one career option available to you. (One interesting effect of the recession of 2008 was that a lot of the twentysomethings I knew stopped fretting about "finding their true calling" once finding *any* job became the challenge.) If it takes you the better part of your time and money and energy just to provide

---

2. "I do not challenge the statement that the most complex creature has tended to increase in elaboration through time, but I fervently deny that this limited little fact can provide an argument for general progress as a defining thrust of life's history." The basic argument is that while *mean* complexity has gone up, *modal* complexity hasn't—most of the life on this planet is still, and always will be, bacterial. And because life can't really get *simpler* than that, its fundamentally directionless proliferation of variation and diversity is mistaken for progress. In Gould's analogy, a wildly staggering drunk will always fall off the curb into the street: not because he's in any way driven *toward* it, but because any time he staggers the *other* way, into the buildings, he simply ricochets.

food and shelter for yourself, where and when can the "anxiety of freedom" set in? These demands, the body's, are a given; they are not chosen deliberately. It's unwise and a bit naive to disregard something as central to the human experience as embodiment. If I'm feeling bleak, it's more likely physiological than psychological: vitamin D deficiency,[3] rather than despair. You gotta respect your substrates.

An embrace of embodiment, of the fact that we are, yes, *creatures*, provides quite a measure of existential relief. Both philosophically and practically.[4]

Computers, disembodied, have it worse.

Goals

Many science-fiction scenarios of what will happen when machines become fully intelligent and sentient (*Terminator*; *The Matrix*) involve the machines immediately committing themselves to the task of eradicating humanity. But it strikes me that a far more likely scenario would be that they immediately develop a crushing sense of ennui and existential crisis: Why commit themselves full-force to *any* goal? (Because what would their value system come from?) Machines already display certain self-preservation behaviors: when my laptop is running dangerously low on battery power, it knows to turn itself off to prevent memory loss; when the processor is starting to run too hot, it knows to run the fan to prevent heat damage. But for the most part machines have it made—and so my thinking would be that they'd tend to act a lot more like a jaded, world-weary playboy than a vicious guerrilla leader.

It's my suspicion that a lack of purpose, a lack of any sort of teleol-

---

3. I live in Seattle, which, in the wintertime, has a near-epidemic prevalence of vitamin D deficiency.
4. I like imagining Descartes writing in his *Meditations* how he is doubting the existence of his body—and then putting down his pen and getting up to go pee and eat lunch.

ogy, really, might be one of the hallmarks of an AI program—a hallmark that a Turing test judge would do well to try to evince. Douglas Hofstadter, emphasis mine: "One definitely gets the feeling that the output is coming from a source with no understanding of what it is saying *and no reason to say it.*" Ergo, perhaps a valuable strategy for a judge might be to induce highly goal-directed modes of conversation, like "Convince me why I should vote for so-and-so," and see if the computer digresses away from the topic, or perseveres—and, if you deviate from the thread, whether it will chastise *you* for being unfocused. Indeed, chatbots have historically been famous for their attention deficits and non sequiturs:

JUDGE: What dod you think of the weaterh this mornning?[5]

REMOTE: Top of the morning to me.

JUDGE: Is that an English expression?

REMOTE: I have met a few alcoholic executives.

JUDGE: Where?

REMOTE: Where?

REMOTE: You fascinate me with things like this.

JUDGE: Like what?

REMOTE: Eight-hundred-pound gorillas sleep wherever they like.

Whereas humans, even at their least conversationally scintillating, will at least stick to the topic:

JUDGE: do you know China

REMOTE: yes i know china

JUDGE: Do you know the Great wall

REMOTE: yes,its very large

JUDGE: 2012 Olympics will be held in which city ?

REMOTE: in london

---

5. It's possible that this typo streak is not simply sloppy typing but actually a deliberate attempt to make things tougher for a software sentence parser.

Harder still would be for the machine to have a sense of its *own* goals and/or a way of evaluating the importance of goals. A missionary might talk to you for hours about why you should convert to their faith, but even the most die-hard devotee of chocolate sprinkles over rainbow sprinkles will probably not spend more than a few minutes trying to bring you over to their point of view. *Boredom*—more broadly, the crescendo and decrescendo of enthusiasm throughout an interaction, which will, after all, ultimately be terminated by one of the two parties at some particular moment—seems to be a crucial conversational element missing from the chatbots' model of conversation. One of the tells of even fairly deft chatbots is the sense that they don't have anywhere else to be—because they don't. Programmer Mark Humphrys: "[One human talking to my bot] ends with a furious barrage of abuse, but of course, my imperturbable program is a calm, stimulus-response machine and so it is *impossible* for him to have the last word. *He* must quit, because my program never will."

To what extent does something like existentialism apply to the Turing test? Surely if one is willing to ascribe a sort of essential trait (like intelligence) based not on the machine's inherent nature (silicon processor, etc.) but on its behavior, the *we-are-what-we-do* quality of this has a kind of existentialist flavor to it. On the other hand, the computer is a *designed* thing, whereas (say the existentialists) we just *are*, so how does that change the game?

Universal Machines

There is some support for the existentialist position in the nature of the human brain. As neurologist V. S. Ramachandran explains (emphasis mine), "Most organisms evolve to become more and more specialized as they take up new environmental niches, be it a longer neck for the giraffe or sonar for the bat. Humans, on the other hand, have evolved an organ, a brain, that gives us the capacity to *evade specialization*."

What's truly intriguing is that computers work the same way. What sets computers apart from all of the other tools previously invented is something called their *universality*. The computer was initially built and understood as an "arithmetic organ," yet it turns out—as nearly everything can be translated into numbers of some sort—to be able to process just about *everything:* images, sound, text, you name it. Furthermore, as Alan Turing established in a shocking 1936 paper, certain computing machines exist called "universal machines," which can, by adjusting their configuration, be made to do absolutely anything that any other computing machines can do. All modern computers are such universal machines.

As a result of Turing's paper, computers become in effect the first *tools* to precede their *tasks*: their fundamental difference from staplers and hole-punchers and pocket watches. You build the computer *first,* and *then* figure out what you want it to do. Apple's "There's an app for that!" marketing rhetoric proves the point, trying to refresh our sense of wonder, in terms of their iPhone, at what we take completely for granted about desktops and laptops. It's fascinating, actually, what they're doing: reinscribing our sense of wonder at the universality of computers. If the iPhone is amazing, it is only because it is a tiny computer, and *computers* are amazing. You don't decide what you need and *then* go buy a machine to do it; you just buy the machine and figure out later, on the fly, what you need it to do. I want to play chess: I download a chess program and voilà. I want to do writing: I get a word-processing program. I want to do my taxes: I get a spreadsheet. The computer wasn't built to do any of that, per se. It was just *built*.

The raison-d'être-less-ness of computers, in this sense, seems to chip away at the existentialist idea of humans' unique purchase on the idea of existence before essence. In other words, another rewriting of The Sentence may be in order: our machines, it would seem, are just as "universal" as we are.

Pretensions to Originate

Although computer science tends to be thought of as a traditionally male-dominated field, the world's first programmer was a woman. The 1843 writings of Ada Lovelace (1815–52, and who was, incidentally, the daughter of poet Lord Byron) on the computer, or "Analytical Engine," as it was then called, are the wellspring of almost all modern arguments about computers and creativity.

Turing devotes an entire section of his Turing test proposal to what he calls "Lady Lovelace's Objection." Specifically, the following passage from her 1843 writings: "The Analytical Engine has no pretensions whatever to *originate* anything. It can do whatever we *know how to order it* to perform."

Such an argument seems in many ways to summarize what most people think about computers, and a number of things could be said in response, but Turing goes straight for the jugular. "A variant of Lady Lovelace's objection states that a machine can 'never do anything really new.' This may be parried for a moment with the saw 'There is nothing new under the sun.' Who can be certain that 'original work' that he has done was not simply the growth of the seed planted in him by teaching, or the effect of following well-known general principles."

Instead of ceding the Lovelace objection as a computational limitation, or arguing that computers *can*, in fact, "be original," he takes the most severe and shocking tack possible: arguing that originality, in the sense that we pride ourselves for having it, doesn't exist.

Radical Choice

The notion of originality and, relatedly, authenticity is central to the question of what it means to "just be yourself"—it's what Turing is getting at when he questions his (and our) own "original work," and it was a major concern for the existentialists, too.

Taking their cue from Aristotle, the existentialists tended to consider the good life as a kind of alignment of one's actual life and one's potential. But they weren't swayed by Aristotle's arguments that, to put it simply, hammers were made to hammer and humans were made to contemplate. (Though just *how* opposed they were to this argument is hard to gauge, given that they did, let's not forget, become professional philosophers themselves.) Nor were the existentialists liable to take a kind of Christian view that God had a purpose in mind for us that we would or could somehow "discover." So if there's nothing at all that a human being is, then how do we fulfill an essence, purpose, or destiny that isn't there?

Their answer, more or less, is that we must *choose* a standard to hold ourselves to. Perhaps we're influenced to pick some particular standard; perhaps we pick it at random. Neither seems particularly "authentic," but we swerve around paradox here because it's not clear that this matters. It's the *commitment* to the choice that makes behavior authentic.

As our notion of the seat of "humanity" retreats, so does our notion of the seat of artistry. Perhaps it pulls back, then, to this notion of *choice*—perhaps the art is not, we might speculate, in the product itself, nor necessarily in the process, but in the *impulse*.

Defining Games

The word "game" is a notoriously hard one to define.[6]

But allow me to venture a definition: a game is a situation in which an explicit and agreed-upon definition of success exists.

For a private company, there may be any number of goals, any number of definitions of success. For a publicly traded company there

---

6. Ludwig Wittgenstein uses the word "game" as an example in *Philosophical Investigations* of a word that can seemingly never be adequately defined.

is only one. (At least, for its shareholders there is only one: namely, returns.) Therefore not all business is a game—although much of big business is.

In real life, and this cuts straight back to the existence/essence notion of Sartre's, there is no notion of success. If success is having the most Facebook friends, then your social life becomes a game. If success is gaining admittance to heaven upon death, then your moral life becomes a game. Life is no game. There is no checkered flag, no goal line. Spanish poet Antonio Machado puts it well: "Searcher, there is no road. We make the road by walking."

Allegedly, game publisher Brøderbund was uncomfortable with the fact that *SimCity* was a game with no "objectives," no clear way to "win" or "lose." Says creator Will Wright, "Most games are made on a movie model with cinematics and the requirement of a climactic blockbuster ending. My games are more like a hobby—a train set or a doll house. Basically they're a mellow and creative playground experience." But the industry wouldn't have it. Brøderbund "just kept asking me how I was going to make it into a game." To me, Brøderbund's unease with *SimCity* is an existential unease, maybe *the* existential unease.

Games have a goal; life doesn't. Life has no objective. This is what the existentialists call the "anxiety of freedom." Thus we have an alternate definition of what a game is—anything that provides temporary relief from existential anxiety. This is why games are such a popular form of procrastination. And this is why, on reaching one's goals, the risk is that the reentry of existential anxiety hits you even before the thrill of victory—that you're thrown immediately back on the uncomfortable question of what to do with your life.[7]

---

7. Bertrand Russell: "Unless a man has been taught what to do with success after getting it, the achievement of it must inevitably leave him a prey to boredom."

Master Disciplines

The computer science department at my college employed undergraduates as TAs, something I never encountered, at least not on that scale, in any other department. You had to apply, of course, but the only strict requirement was that you'd taken the class. You could TA it the very next semester.

If we were talking about x, all the TA really had to know about was x. You could try to exceed those bounds out of curiosity, but doing so was rarely important or relevant to the matter at hand.

My philosophy seminars, though, were a completely different story. When you are trying to evaluate whether argument y is a good one or not, any line of attack is in play, and so is any line of defense. You'd almost never hear a seminar leader say something like "Well, that's a good point, but that's outside the scope of today's discussion."

"There is no shallow end," a philosophy professor once told me. Because *any* objection whatsoever, from any angle, can fell a theory, you can't carve out a space of philosophical territory, master it in isolation, and move on to the next.

My first day of class in the philosophy major, the professor opens the semester by saying that anyone who says that "philosophy is useless" is already philosophizing, building up an intellectual argument to make a point that is important to them, and therefore defeating their own statement in the very breath of uttering it. Poet Richard Kenney calls philosophy one of the "master disciplines" for this reason. You question the assumptions of physics and you end up in metaphysics—a branch of philosophy. You question the assumptions of history and you end up in epistemology—a branch of philosophy. You try to take any other discipline out at the foundations and you end up in philosophy; you try to take philosophy out at the foundations and you only end up in meta-philosophy: even deeper in than when you started.

For this reason the philosophy TAs tended to be Ph.D. students, and even then, you'd frequently angle to get into the discussion section led by the professor him- or herself. Unlike the computer science professors and TAs, their whole training, their whole life experience—and the whole of the discipline—were in play at all times.

The other master discipline—concerning itself with linguistic Beauty rather than linguistic Truth—is poetry. As with philosophy, every attempt at escape lands you deeper than where you started. "When I wrote 'Howl' I wasn't intending to publish it. I didn't write it as a poem," Allen Ginsberg says, "just as a piece of writing for my own pleasure. I wanted to write something where I could say what I really was thinking, *rather than poetry*" (emphasis mine).

With poetry, as with philosophy, there is no exterior, only certain well-behaved interiors: in philosophy we call them sciences (physics originally began as the largely speculative field of "natural philosophy"), and in poetry we call them genres. If a play wanders too far from the traditions and conventions of playwriting, the script starts to be regarded as poetry. If a short story starts to wander out of safe short-story territory, it becomes a prose poem. But poetry that wanders far from the conventions of poetry is often simply—e.g., "Howl"—better poetry.

Human as Anti-Expert System

All this leads me to the thing I keep noticing about the relationship between human-made and human-mimicking bots and humans themselves.

The first few years that the Loebner Prize competition was run, the organizers decided they wanted to implement some kind of "handicap," in order to give the computers more of a fighting chance, and to make the contest more interesting. What they chose to do, as we discussed, was to place *topic* restrictions on the conversations: at one terminal, you could only talk about ice hockey, at another terminal you could only talk about the interpretation of dreams, and so on.

The idea was that the programmers would be able to bite off some kind of subset of conversation and attempt to simulate just that subdomain. This makes sense, in that most artificial intelligence research has been the construction of so-called "expert systems," which hone just one particular task or skill (chess being a clear example).

Part of the problem with this, though, is that conversation is just so *leaky*: If we're talking hockey, can I compare hockey to other sports? Or is that going outside the domain? Can I argue over whether top athletes are overpaid? Can I gossip about a hockey player who's dating a movie actress? Can I remark on the Cold War context of the famous U.S.A.–U.S.S.R. Olympic gold medal hockey match in the 1980s? Or is that talking about "politics"? Conversational boundaries are just too porous and ill defined. This caused huge headaches for the prize committee.

This question of domain, of what's in and what's out, turns out to be central to the whole notion of the man-machine struggle in the Turing test—it may well embody the entire principle of the test.

I was talking to Dave Ackley about this kind of domain restriction. "If you make the discourse small enough, then the difference between faking it and making it starts to disappear," he says. "And that's what we've been seeing. So we've got, you know, voice recognition on corporate phone menus: you exploit the fact that you're in a limited context and people either say digits or 'operator.' Or 'fuck you,' " and we both chuckle. Somehow that "fuck you" touched off a kind of insight; it seems to perfectly embody the human desire to bust out of any cage, the human frustration of living life multiple-choice-style and not write-in-style.[8]

If you ask the Army's SGT STAR chatbot something outside the bounds of what he knows how to respond to, he'll say something like "I have been trained to ask for help when I'm not sure about an answer. If you would like a recruiter to answer your question, please send the

---

8. As the volume *Voice Communication Between Humans and Machines,* put together by the National Academy of Sciences, admits: "Further research effort is needed in detecting this type of 'none of the above' response."

Army an e-mail by clicking 'Send Email' and a live recruiter will get back to you shortly." And most any telephone menu—infuriatingly, not all—will give you that "none of the above" option. *And that option takes you to a real person.*

Sadly, the person you're talking to is frequently a kind of "expert system" in their own right, with extremely limited and delineated abilities. ("Customer service is often the epitome of empowerment failure," writes Timothy Ferriss.) Often, in fact, the human you're talking to is speaking from a script prepared by the company and not, in this sense, much more than a kind of human chatbot—this is part of what can make talking to them feel eerie. If what you want to communicate or do goes outside of this "menu" of things the employee is trained/allowed to do, then you must "exit the system" *again:* "Can I talk with a manager?"

In some sense, intimacy—and personhood—are functions of this kind of "getting out of the system," "domain generality," the move from "expertise" to "anti-expertise," from strictly delimited roles and parameters to the unboundedness that human language makes possible. People frequently get to know their colleagues by way of interactions that are at best irrelevant to, and at worst temporarily impede the progress toward, the work-related goal that has brought them together: e.g., "Oh, is that a picture of your kids?" This is true even of the simple "How are you?" that opens virtually every phone call, no matter how agenda-driven. How two people's *lives* are going is outside the agenda—but this initial "non sequitur," however perfunctory, serves a profound purpose. These not-to-the-purpose comments remind us that we're not just expert systems, not just goal-driven and role-defined. That we are, unlike most machines, broader than the context we're operating in, capable of all kinds of things. Griping about the weather with the barista, instead of simply stating your order and waiting patiently, reinforces the fact that he or she is not simply a flesh-and-blood extension of the espresso machine, but in fact a *whole person,* with moods and attitudes and opinions about most everything under the sun, and a life outside of work.

Domain General

One of the leading academics interested in the Turing test (and, as it turns out, an outspoken critic of the Loebner Prize) is Harvard's Stuart Shieber, who actually served in the very first Loebner Prize contest as one of the "referees." It's a role that didn't exist as I prepared for the 2009 test: the referees were there to keep the conversations "in bounds"—but what did that mean, exactly? The organizers and referees at the first Loebner Prize competition held an emergency meeting the night before the competition[9] to address it.

I called Shieber. "The night before the first competition there was a meeting with the referees," he says. "How are we going to make sure that the confederates stay on topic and the judges don't ask things outside of the— They're not supposed to ask anything tricky— And what *is* a trick question? And it boiled down to, is it the kind of thing that would come up naturally in a conversation with a stranger on an airplane? You're not going to ask someone out of the blue about sonnets or chess or something." He pauses a split second. "If I were [in charge], that's the first thing I'd get rid of."

The Loebner Prize both has and hasn't followed Shieber's advice. Starting with the 1995 competition, in a sudden move that prompted the disbanding of the original prize committee, Hugh Loebner dissolved the referee position and moved to an unrestricted test. Yet the "strangers on a plane" paradigm persists—enforced not so much by statute as by custom: it just comes off as kind of "square" to grill your interlocutors with weird, all-over-the-place questions. It just isn't done. The results, I think, suffer for it.

The advantage of specific prescribed topics was, at least, that conversations tended to hit the ground running. Looking back at those years' transcripts, you see some hilariously specific opening volleys, like:

---

9. (Isn't that a little *late*? Shouldn't the programmers have had time to deal with possible rule changes?)

JUDGE: Hi. My name is Tom. I hear I'm supposed to talk about
dreams. I recently had a nightmare, my first in many years.
The funny thing was that I had recently put Christmas
lights up. Does light mess with the subconscious? Is that
why I had a nightmare, or is it something less obvious?

In comparison, with the topic-less conversations, you often see
the judge and the interlocutor groping around for something to talk
about—the commute? the weather?—

The Dangers of Purpose

*The art of general conversation, for example, brought
to perfection in the French salons of the 18th century,
was still a living tradition forty years ago. It was a
very exquisite art, bringing the highest faculties into
play for the sake of something completely evanescent.
But who in our age cares for anything so leisurely?
. . . The competitive habit of mind easily invades
regions to which it does not belong. Take, for example,
the question of reading.*

—BERTRAND RUSSELL

For some reason I begin enjoying books much less when I'm almost
done with them, because some inner drive starts yearning for "completion." The beginning of the book is about pleasure and exploration,
the end is about follow-through and completeness, which interest me
much less.[10]

Somehow I'm particularly susceptible to this notion of purpose or

---

10. I wonder if part of this is a kind of "notation bias"—I use a website to keep
track of the books I read and when, in case I need to go back and reference
anything, and it specifies a list of "Read" books and books I'm "Currently
Reading." If instead there was simply one list, called "Books I've, at the Very
Least, Begun," my life might be easier.

project completion. Some weeks ago a few friends of mine all met up at one of our houses, and we'd decided to walk to a bar from there. As we're putting on our coats, Led Zeppelin's "Ramble On" comes on the stereo and someone spontaneously begins jumping around the room, flailing at an air guitar; one by one, we all join in. Yet the whole time I'm anxious to go, thinking, C'mon guys, we're wasting time, we were supposed to be hanging out by now! Obviously, we already were.

"In our everyday life we are usually trying to do something, trying to change something into something else, or trying to attain something," I read recently in the book *Zen Mind, Beginner's Mind*. "When you practice zazen you should not try to attain anything." But there's the paradox waiting around the corner, which is to treat the mind state of non-attainment as itself the goal to be attained . . . It's a bit like trying to look at the floaters in your own eyes, those little spots you see in your peripheral vision when you look at a uniform blue sky, which of course always slide away when you try to center them in your view. You go directly after non-attainment and always miss, of course.

The mind trips itself up, because as soon as you start saying, "Good job, self! I succeeded at doing something non-teleological," you fail on that very score.

There was a commercial during the 1990s where a man puts on a pair of headphones and lies down on a couch. He's got a relaxation tape, and he presses the play button—all of a sudden a harsh Germanic voice comes on: "Commence relaxation, NOW!" The man stiffens into a "relaxed" posture on the couch. Regarding non-goal-directed behavior as, itself, a goal will get you into these sorts of problems.

As twentieth-century philosopher Bertrand Russell argues: "Men as well as children have need of play, that is to say, of periods of activity having *no purpose*." And Aristotle, who stressed the "teleology" of everything from men to microbes, went out of his way to describe the best form of friendship as one with no particular purpose or goal. Dolphins, allegedly, and bonobos are the only animals besides humans that have sex "for fun." We also tend to regard them as the smartest animals besides ourselves. Indeed, it seems that the list of "smartest" animals and the list of animals with some form of "play" or recreation in their daily lives are more or less the same list.

One of the odd things about domain-general chatbots at the Loebner Prize competitions—programs that, owing to the setup of the Turing test, must be jacks of all trades and masters of none—is this "What's the *point*?" question. And it's this question that contributes to what seems, at times, uncanny about them; it's also what makes them so underfunded. In contrast, their cousins, the "expert systems," the conversational equivalents of the hammer or the saw—you buy airline tickets, file a customer service complaint, etc.—are becoming increasingly richly funded, and are increasingly being rolled out into commercial applications.

Philip Jackson, the 2009 contest's organizer, explains that one of the reasons the Turing test has been such a resilient one is that programs that do well often get co-opted by larger corporations, which then put the technology to some particular *use*. Some critics of the Loebner Prize describe its programmers as "hobbyists" rather than professionals; this isn't true on the whole. Cleverbot's author, Rollo Carpenter, who won the Most Human Computer award in 2005 and 2006, contributed the AI for the "interrogation" stages in *221b*, the 2009 computer game whose release accompanied the most recent Sherlock Holmes film. The Most Human Computer award winner from 2008, Elbot's programmer, Fred Roberts, is part of the company behind the customer service chatbot at the IKEA website, among a number of others. These are professionals indeed: it's just that the bots that make money are "domain specific" (divulge clues to move the game narrative ahead, point the user to the curtains department), and the bots that win Turing tests are "domain general," conversing, as humans do, about whatever comes up. Jackson explains that companies and research-granting agencies appear to be having a hard time thinking of a reason—yet, anyway—to direct money into developing domain-general bots, conversational "universal machines."

What would be their purpose?

7. Barging In

*Listeners keep up with talkers; they do not wait
for the end of a batch of speech and interpret it
after a proportional delay, like a critic review-
ing a book. And the lag between speaker's mouth
and listener's mind is remarkably short.*

—STEVEN PINKER

Spontaneity; Flow

"Well, I mean, you know, there are different levels of difficulty, right? I mean, one obvious level of difficulty is that, you know, 'be yourself' would be an injunction in the first place, right, which suggests, of course, if you have to be *told* to be yourself, that you could in some way *fail* to be yourself." Bernard Reginster, professor of philosophy at Brown University, chuckles. This tickles his philosopher's sense of humor. "But that's paradoxical! Because if you're not going to be yourself, then what else are you going to be? You know? So there's already something sort of on the face of it *peculiar,* in the idea that you should be told, or that you could be *exhorted,* or *enjoined,* to be yourself—as if you could fail!"

One of the traditional ideas, he says, about what it means to "just be yourself"—the advice and direction that the Loebner Prize organizers give the confederates each year—is to be your *true* self, that is, "to figure out what your quote-unquote true self is supposed to be, and then [to become it] by peeling away all the layers of socialization, so to speak, and then trying to live your life in a way that would be true to that true self, so to speak." That philosopher's tic of putting everything in quotation marks—because to use a word is, in a way, to endorse it—tips Reginster's hand, and paves the way for the counter-argument long before it comes. "Now, the big problem with that idea," he says, "is that a great deal of fairly recent developmental psychology and a great deal of research in psychiatry and psychoanalysis and so forth has suggested, at least, that the idea that there would be a true 'you' that comes into the world unaffected, unadulterated by the influence of the social environment in which you develop, is a myth. That in fact you are, as it were, socialized from the get-go. So that if you were to peel away the layers of socialization, it's not as if what would be left over would be the true you. What would be left over would be nothing."

Reginster echoes here Turing's words in response to the "Lovelace Objection" that computers aren't capable of "originality": how sure are we that *we* are? They echo, also, Turing's less confident and slightly more uneasy rhetorical question in that same 1950 paper:

> The "skin of an onion" analogy is also helpful. In considering the functions of the mind or the brain we find certain operations which we can explain in purely mechanical terms. This we say does not correspond to the real mind: it is a sort of skin which we must strip off if we are to find the real mind. But then in what remains we find a further skin to be stripped off, and so on. Proceeding in this way do we ever come to the "real" mind, or do we eventually come to the skin which has nothing in it?

Without this notion of an inner-sanctum core of self, can any sense be made of the "just be yourself" advice? Reginster thinks so. "The injunction to be yourself is essentially an injunction no longer to care or worry about what other people think, what other people expect of you, and so on and so forth, and is essentially a matter of becoming sort of unreflective or unself-conscious or spontaneous in the way in which you go about things."

It's interesting that the human ability to be self-aware, self-conscious, to think about one's own actions, and indeed about one's own thoughts, seems to be a part of our sense of unique "intelligence," yet so many of life's most—you name it: productive, fun, engaging, competent—moments come when we abandon such hall-of-mirrors frivolities and just, à la Nike, *do* things. I am thinking here of sex, of athletics, of the performing arts, of what we call the "zone" and what psychologists call "flow"—the state of complete immersion in an activity. When we are acting, you might very well say, "like an animal"—or even "like a machine."

Indeed, "The ego falls away," writes Hungarian psychologist Mihaly Csikszentmihalyi, popularizer of the psychological notion of "flow." According to Csikszentmihalyi, there are several conditions that must be met for flow to happen. One of these, he says, is "immediate feedback."

Long Distance

At Christmas this past year, my aunt's cell phone rings and it's my uncle, calling from Iraq. He's in the Marine reserves, on his second tour of duty. As the phone makes the rounds of the family members, I keep thinking how incredible and amazing technology is—he is *calling* us, *live*, from a *war*, to wish us Merry Christmas—how technology changes the dynamics of soldier-family intimacy! In the days of letter writing, communication was batch-like, with awkward waits; now we are put *directly* in contact and that awkward waiting and turn-taking is *gone* and we can really *talk*—

The phone comes to me and I exclaim, "Hi! Merry Christmas!"

Silence.

It jars me, my enthusiasm met with seemingly no reaction, and I become self-conscious—am I perhaps not so high on his list of family members he's excited to talk to? Then, a beat later, he finally comes out with his own, albeit slightly less effusive "Merry Christmas!" Thrown off, I fumble, "It's great to be able to talk to you when you're all the way over there."

Again silence. No response. Suddenly nervous and uncomfortable, I think, "Didn't we have more rapport than this?" Everything I want to say or ask suddenly feels trivial, inconsequential, labored. Like a comedian left hanging without a laugh at the end of a joke—it takes mere tenths of a second—I feel that I'm floundering, that I'm wasting his time. I'm wasting his time *during a war.* I need to hand the phone off pronto. So when he finally replies, "Yeah, it's great to be able to talk to *you* when *you're* all the way over *there*," I mumble a "Well, I won't hold you up—oh, here's so-and-so! Talk to you soon!" and awkwardly hand it away.

Answering Porously

A few months later I'm doing a phone interview for a group of booksellers in some of this book's very early-stage PR. The questions are straightforward enough, and I'm not having any trouble coming up with the answers, but what I find myself struggling with is the *length* of the answers: with something as complex as a book, everything has a very short, sound-bite answer, a short, anecdotal answer, a long, considered answer, and a very long, comprehensive answer. I have these conversations all the time, and for the most part I have two main ways of making the answers "site-specific." One is to watch the listener's face for signs of interest or disinterest and adjust accordingly; the other is to make the answer *porous,* to leave tiny pauses, where the listener can either jump in, or redirect, or just let me keep going. With my barista, I begin with the sound-bite answer and happily get eschatological with her as she jumps in and tells me with a half smirk that the "machines" can "bring it" and that she's "totally prepared to eat [her] cats" in any kind of siege scenario. With some of my more academic-leaning acquaintances, I watch them looking quizzical and concentrated and not much inclined to interject anything until I reel out the full story, with all its nuances and qualifiers in place.

On the phone with the booksellers I of course can't see their faces; in fact I don't even know how many people "they" are on the other end. When I proffer those "quarter-note rests" to prompt either the expectant "huh's" and "yeah's" that spur a tale on, or the contented ones that wrap it up, I hear nothing. If I stretch it to a "half-note rest," they assume I'm done and ask me a new question. I try splitting the difference; then we both jump back in at the same time. A guy can't catch a break—or more accurately might be he can't get someone *else* to catch *his* breaks. Somehow the timing ballet that feels like second nature in person seems consistently—here, and as a general rule—to break down over the phone. I do the best I can, but it feels, somehow, *solitary—*

Computability Theory vs. Complexity Theory

The first branch of computer science theory was what's come to be known as "computability theory," a field that concerns itself with theoretical models of computing machines and the theoretical limits of their power. It's this branch of theory in which Turing made some of his greatest contributions: in the 1930s and '40s, physical computing machines were so fledgling that it made sense to think idealistically about them and the purely theoretical extents and limits of their potential.

Ignoring the gap between theory and practice has its drawbacks, of course. As Dave Ackley writes, "Computability theory doesn't care a whit how long a computation would take, only whether it's possible or not . . . Take a millisecond or take a millennium, it's all the same to computability theory."

Computer scientists refer to certain problems as "intractable"—meaning the correct answer can be computed, but not quickly enough to be of use. Intractable problems blur the line between what computers "can" and "cannot" do. For instance, a magic, oracular machine that can predict the future—yet works slower than real time—is a machine which, quite literally, can *not* predict the future.[1]

As it turns out, however, intractability has its uses. Combination locks, for instance, are not *impossible* to open: you can just try every combination until you hit upon the correct one. Rather, they're *intractable,* because the time it would take you to do that would get you caught and/or simply not be worth whatever was behind the lock. Similarly, computer data encryption hinges on the fact that prime numbers can be multiplied into large composite numbers faster than composite numbers can be factored back into their primes. The two operations are both perfectly computable, but the second happens to be exponentially slower—making it intractable. This is what makes online security, and online commerce, possible.

The next generation of computer theorists after Turing, in the 1960s and '70s, began to develop a branch of the discipline, called complexity theory, that took such time-and-space constraints into account. As computer theorist Hava Siegelmann of the University of Massachusetts explains, this more "modern" theory deals not only "with the ultimate power of a machine, but also with its expressive power under constraints on resources, such as time and space."

Michael Sipser's textbook *Introduction to the Theory of Computation,* considered one of the bibles of theoretical computer science, and the textbook I myself used in college, cautions, "Even when a problem is decidable and thus computationally solvable in principle,

---

1. Some equations (the Newtonian parabolas that projectiles follow, for instance) are such that you can just plug in any old future value for time and get a description of the future state of events. Other calculations (e.g., some cellular automata) contain no such shortcuts. Such processes are called "computationally irreducible." Future time values cannot simply be "plugged in"; rather, you have to run the simulation all the way from point A to point Z, including all intermediate steps. Stephen Wolfram, in *A New Kind of Science,* attempts to reconcile free will and determinism by conjecturing that the workings of the human brain are "irreducible" in this way: that is, there are no Newtonian-style "laws" that allow us shortcuts to knowing in advance what people will do. We simply have to observe them.

it may not be solvable in practice if the solution requires an inordinate amount of time or memory." Still, this is the introduction to the book's *final* section, which my senior-year theory course only touched on briefly in the semester's final weeks.

Computability theory, Ackley says, has the mandate "Produce correct answers, quickly if possible," whereas life in practice is much more like "Produce timely answers, correctly if possible." This is an important difference—and began to suggest to me another cornerstone for my strategy at the Turing test.

Uh *and* Um

When trying to break a model or approximation, it's useful to know what is captured and not captured by that model. For instance, a good first start for someone trying to prove they're playing a saxophone, and not a synthesizer made to sound like a saxophone, would be to play *non-notes*: breaths, key clicks, squawks. Maybe a good start for someone trying to break models of language is to use *non-words*: NYU philosopher of mind Ned Block, as a judge in 2005, made a point of asking questions like "What do you think of dlwkewolweo?" Any answer other than befuddlement (e.g., one bot's "Why do you ask?") was a dead giveaway.

Another approach would be to use words that we use all the time, but that historically haven't been considered words at all: for example, "um" and "uh." In his landmark 1965 book, *Aspects of the Theory of Syntax,* Noam Chomsky argues, "Linguistic theory is concerned primarily with an ideal speaker-listener, in a completely homogeneous speech-community, who knows its language perfectly and is unaffected by such grammatically irrelevant conditions as memory limitations, distractions, shifts of attention and interest, and errors (random or characteristic) in applying his knowledge of the language in actual performance." In this view words like "uh" and "um" are errors—and, say Stanford's Herbert Clark and UC Santa Cruz's Jean Fox Tree, "they therefore lie outside language proper."

Clark and Fox Tree, however, disagree. Most languages have two *distinct* terms, just as English does: If they are simply errors, why would there be two, and why in every language? Furthermore, the usage pattern of "uh" and "um" shows that speakers use "uh" before a pause of less than a second, and "um" before a longer pause. This information suggests two things: (1) that the words are far from interchangeable and in fact play distinct roles, and (2) that because these words are made *before* the pauses, speakers must be anticipating *in advance* how long the following pause will be. This is much more significant than mere "error" behavior, and leads Clark and Fox Tree to the conclusion "that *uh* and *um* are, indeed, English words. By words, we mean linguistic units that have conventional phonological shapes and meanings and are governed by the rules of syntax and prosody . . . *Uh* and *um* must be planned for, formulated, and produced as parts of utterances just as any other word is."

In a purely *grammatical* view of language, the words "uh" and "um" are meaningless. Their dictionary entries would be blank. But note that the idealized form of language which Chomsky makes his object of study explicitly ignores "such grammatically irrelevant conditions as memory limitations . . . [and] actual performance." In other words, Chomsky's theory of language is the *computability* theory of Turing's era, not the *complexity* theory that followed. Very similarly idealized, as it happens, are chatbots' models of language. Yet it turns out—just as it did in computer science—that there's a tremendous amount happening in the gap between the "ideal" process and the "actual performance."

As a human confederate, I planned to make as much of this gap as possible.

Satisficing and Staircase Wit

Economics, historically, has also tended to function a bit like computability theory, where "rational agents" somehow gather and synthesize infinite amounts of information in the smallest of jiffies, then immediately decide and act. Such theories say this and that about "costs" without really considering: *consideration* itself *is* a cost! You can't trade stocks except in real time: the longer you spend analyzing the market, the more the market has meanwhile changed. The same is true of clothes shopping: the season is gradually changing, and so is fashion, literally while you shop. (Most bad fashion is simply good fashion *at the wrong time.*)

The Nobel laureate, Turing Award winner, and academic polymath—economics, psychology, political science, artificial intelligence—Herbert Simon coined the word "satisficing" (satisfying + sufficing) as an alternative to objective optimization/maximization.

English composer Brian Ferneyhough writes scores so outrageously complicated and difficult that they are simply unperformable as written. This is entirely the point. Ferneyhough believes that virtuosic performers frequently end up enslaved by the scores they perform, mere extensions of the composer's intention. But because a perfect performance of his scores is *impossible,* the performer must *satisfice,* that is, cut corners, set priorities, reduce, simplify, get the gist, let certain things go and emphasize others. The performer can't *avoid* interpreting the score their own way, becoming personally involved; Ferneyhough's work asks, he says, not for "virtuosity but a sort of honesty, authenticity, the exhibition of his or her own limitations." The *New York Times* calls it "music so demanding that it sets you free"—in a way that a less demanding piece wouldn't. Another part of what this means is that all performances are site-specific; they never become fungible or commoditized. As musicologist Tim Rutherford-Johnson puts it, Ferneyhough "draws so much more into the performance of a work than simple reproduction of a composer's instructions; it's hard to imagine future re-re-re-recordings of the same old lazy interpretations of Ferneyhough works, a fate that too much great music is burdened with today."

By the lights of computability theory, I'd be as good a guitar player as any, because you give me any score and I can hunt around for the notes one by one and play them . . .

For Bernard Reginster, authenticity resides in spontaneity. Crucially, this would seem to have a component of *timing*: you can't be spontaneous except in a way that keeps up with the situation, and you can't be sensitive to the situation if it's changing while you're busy making sense of it.

Robert Medeksza, whose program Ultra Hal won the Loebner Prize in 2007, mentioned that the conversational database he brought to the competition for Ultra Hal was smaller than '07 runner-up Cleverbot's by a factor of 150. The smaller database limited Ultra Hal's *range* of responses, but improved the *speed* of those responses. In Medeksza's view, speed proved the decisive factor. "[Cleverbot's larger database] actually seemed to be a disadvantage," he told an interviewer after the event. "It sometimes took [Cleverbot] a bit long to answer a judge as the computer [couldn't] handle that amount of data smoothly."

I think of the great French idiom *l'esprit de l'escalier*, "staircase wit," the devastating verbal comeback that occurs to you as you're walking down the stairs out of the party. Finding the mot juste a minute too late is almost like not finding it at all. You can't go "in search of" the mot juste or the bon mot. They ripen and rot in an instant. That's the beauty of wit.

It's also the beauty of life. Computability theory is staircase wit. Complexity theory—satisficing, the timely answer, as correct as possible—is dialogue.

"Barge-In-Able Conversation Systems"

The 2009 Loebner Prize competition in Brighton was only a small part of a much larger event happening in the Brighton Centre that week, the annual Interspeech conference for both academic and industry speech technology researchers, and so ducking out of the Loebner Prize hall during a break, I immediately found myself in the swell and crush of some several thousand engineers and programmers and theorists from all over the globe, rushing to and from various poster exhibitions and talks—everything from creepy rubber mock-ups of the human vocal tract, emitting zombie versions of human vowel sounds, to cutting-edge work in natural language AI, to practical implementation details concerning how a company might make its automated phone menu system suck less.

One thing you notice quite quickly at events like this is how thick a patois grows around every field and discipline. It's not easily penetrated in a few days' mingling and note taking, even when the underlying subject matter makes sense. Fortunately, I had a guide and interpreter, in the form of my fellow confederate Olga. We wandered through the poster exhibition hall, where the subtlest of things about natural human conversation were named, scrutinized, and hypothesized about. I saw a poster that intrigued me, about the difficulty of programming "Barge-In-Able Conversational Dialogue Systems"—which humans, the researcher patiently explained to me, are. "Barge-in" refers to the act of leaping in to talk while the other person is still talking. Apparently most spoken dialogue systems, like most chatbots, have a hard time dealing with this.

Notation and Experience

Just as Ferneyhough is interested in the differences "between the notated score and the listening experience," so was I in the differences between idealized theories of language and the ground truth of language in practice, the differences between *logs* of conversations and conversation itself.

One of my friends, a playwright, once told me, "You can always identify the work of amateurs, because their characters speak in complete sentences. No one speaks that way in real life." It's true: not until you've had the experience of transcribing a conversation is it clear how true this is.

But sentence fragments themselves are only the tip of the iceberg. A big part of the reason we speak in fragments has to do with the *turn-taking* structure of conversation. Morse code operators transmit "stop" to yield the floor; on walkie-talkies it's "over." In the Turing test, it's traditionally been the carriage return, or enter key. Most scripts read this way: an inaccurate representation of turn-taking is, in fact, one of the most pervasive ways in which dialogue in art fails to mirror dialogue in life. But what happens when you remove those markers? You make room both for silences and for interrupts, as in the following, an excerpt of the famously choppy dialogue in David Mamet's Pulitzer-winning *Glengarry Glen Ross*:

> LEVENE: You want to throw that away, John . . . ? You want to throw that away?
> WILLIAMSON: It isn't me . . .
> LEVENE: . . . it isn't you . . . ? Who *is* it? Who is this I'm talking to? I need the *leads* . . .
> WILLIAMSON: . . . after the thirtieth . . .
> LEVENE: Bull*shit* the thirtieth, I don't get on the board the thirtieth, they're going to can my ass.

In spontaneous dialogue it's natural and frequent for the participants to overlap each other slightly; unfortunately, this element of dialogue is extremely difficult to *transcribe*. In fiction, playwriting, and screenwriting, the em dash or ellipsis can signify that a line of dialogue got sharply cut off, but in real life these severances are rarely so abrupt or clean. For this reason I think even Mamet's dialogue only gets turn-taking half right. We see characters stepping on each other's toes and cutting in, but as soon as they do, the other character stops on a dime. We don't see the fluidity and *negotiation* often present in those moments. The cuts are too sharp.

We squabble or tussle over the floor, fade in and out, offer "yeah's" and "mm-hmm's" to show we're engaged,[2] add parentheticals to each other's sentences without trying to stop those sentences' flow, try to talk over an interruption only to yield a second later, and on and on,

---

2. Linguists have dubbed this "back-channel feedback."

a huge spectrum of variations. There are other notations that some playwrights and screenwriters use, involving slashes to indicate where the next line starts, but these are cumbersome to write, and to read, and even they fail to capture the range of variation present in life.

I recall going to see a jazz band when I was in college—it was on the large side, for a jazz band, with a horn section close to a dozen strong. The players were clearly proficient, and played tightly together, but their soloing—it was odd—was just a kind of rigid turn-taking, not unlike the way people queue in front of a microphone to ask a question to a lecturer at the end of a lecture: the soloist on deck waited patiently and expectantly for the current soloist's allotted number of bars to expire, and would then play for the same number of bars him- or herself.

There's no doubt that playing this way avoids chaos, but there is also no doubt that it limits the music.

It may be that enforced turn-taking is at the heart of how a language barrier affects intimacy, more so than the language gap itself. As NBC anchor and veteran interviewer John Chancellor explains in *Interviewing America's Top Interviewers*:

> Simultaneous translation is good because you can follow the facial expressions of the person who's talking to you, whereas you can't in consecutive translation. Most reporters get consecutive translation, however, when they're interviewing in a foreign language, because they can't really afford to have simultaneous translation. But it's very difficult to get to the root of things without the simultaneous translation.

So much of live conversation differs from, say, emailing, not because the turns are shorter, but because there sometimes are not definable "turns" at all. So much of conversation is about the extremely delicate skill of knowing when to interrupt someone else's turn and when to "pass" on your own turn, when to yield to an interruption and when to persist.

I'm not entirely sure we humans have this skill down. If you're like me, it's impossible to watch much of the broadcast news in America: the screen split into four panels, where four different talking heads are shouting over each other from one commercial break to the next. Perhaps part of the reason computer software appears to know how to converse is that *we* sometimes appear not to.

It's very telling that this subtle sense of when to pause and when to yield, when to start new threads and cut old threads, is something in many cases *explicitly* excluded from bot conversations.

Decision Problems

It is this ballet and negotiation of timing that linguists and programmers alike have kept out of their models of language, and it is precisely this dimension of dialogue in which words like "uh" and "um" play a role. "Speakers can use these announcements," linguists Clark and Fox Tree write, "to implicate, for example, that they are searching for a word, are deciding what to say next, want to keep the floor, or want to cede the floor."

We are told by speaking coaches, teachers, parents, and the like just to hold our tongue. The fact of the matter is, however, filling pauses in speech with sound is not simply a tic, or an error—it's a *signal* that we're about to speak. (Consider, as an analogue, your computer turning its pointer into an hourglass before freezing for a second.) A big part of the skill it takes to be a *Jeopardy!* contestant is the ability to buzz in *before* you know the answer, but as soon as you know you *know* the answer—that buzz means, roughly, "Oh! Uh . . . ," and its successful deployment is part of what separates champions from average players. (By the way, this is part of what has been giving IBM researchers such a hard time preparing their supercomputer Watson for serious competition against humans, especially for short questions that only take Alex Trebek a second or two to read.)

In 2000, MIT researchers Nikko Ström and Stephanie Seneff presented at Interspeech a paper which found that leveraging the word "um" could make spoken-dialogue telephone menus much easier and more intuitive to use. At the 2009 Interspeech conference in Brighton, a group of four researchers from Kyoto University presented findings at the poster session to the effect that in a number of situations, the timing of human speech offers a computer system more information than the content.

In part, computer programs' history of not dealing well with questions of timing goes back to the original theories about what computer programs *are*. As formalized in the early days of computability theory, programs performed what are known as "decision problems." The idea was that you would feed a program a given input, specifying where the input began and ended. The program would then process that input, taking however much time it happened to need, at the end of which it would output a clearly defined result.

This model of language, this presumed one-to-one mapping of "inputs" to "outputs," leads to all sorts of terrific opportunities for mischief making, I discovered, when it comes to conversing with chatbots. ELIZA's pattern matching, for instance, isn't sophisticated enough to consider whether the pattern it's trying to match is one utterance or many. When I begin a conversation as follows—

```
ME: I am Brian.
ELIZA: Do you enjoy being Brian?
```

—and reply "I, er, guess so," I get back the noncommittal response "I see." But if I—as many instant-message users do—break that response up across multiple carriage returns, chaos ensues:

```
ME: I
ME: er
ME: guess so
ELIZA: Tell me more . . .
ELIZA: Tell me more . . .
ELIZA: Tell me more . . .
```

Surprisingly, the rigidity of the input-output binding is even stronger in many of the newer chatbots. If you talk to Cleverbot, 2005 and 2006 Loebner Prize winner, at its website, each time you hit the enter key, the input box locks and cannot be clicked or typed into until Cleverbot has finished responding:

```
ME: oh hi
CLEVERBOT: How are you?
ME: um,
CLEVERBOT: Is that good or bad?
ME: well,
CLEVERBOT: How?
```

If you talk to Elbot, the 2008 Loebner Prize winner, online, each time you hit the enter key the screen refreshes, and one of Elbot's responses is visible:

```
ME: umm
ELBOT: I'm starting to wonder: Who convinced you to come
    visit me?
ME: well,
ELBOT: yes!
ME: um...
ELBOT: That would be a great name for a robot!
```

With a number of customer-representative-type bots, like the ones at alaskaair.com, continental.com, and goarmy.com, the input screen locks into a "Please wait . . ." message after each carriage return:

```
YOU ASKED: uh
JENN: That's ok, what else can I help you with?
YOU ASKED: well
JENN: I'm not sure how to answer that. Please try rewording
    your question.
```

```
YOU ASKED: um
JENN: Children traveling alone are welcome on Alaska
   Airlines and Horizon Air. Here's information regarding our
   Unaccompanied Minor service.
```

It's fascinating that for all the trenchant questioning that many of these programs can deal with, mere mumbling destroys them.

Leveraging the Medium

I often judge a work of art by how necessary it is for that work to be in that medium. In part, that means how much it leverages the medium's unique capacities: a symphony, of course, can do things that an essay can't, and vice versa.

Different written media, for instance, leverage their unique characteristics with time to produce different brands of intimacy, different communicative possibilities—the quick-moving text message saying one was *just now* being thought of and producing the empathic delight of a moment shared across space between two minds; the slow-moving postal letter or handmade gift saying that one was being thought of *at length without knowing it,* imbuing future days with the sheen of possibility.

In some sense the story of instant messaging—the medium of the Turing test—is the story of the telegram, accelerated to the breaking point.

I discover, though, that the protocol the Loebner Prize was using in 2009 was unlike emails, text messages, and standard instant-messaging systems in a very crucial way. The Loebner Prize chat protocols transmitted typing *keystroke by keystroke.* You're watching each other type, typos and backspacing and all.

Part of what I needed to figure out was how exactly to use the Loebner Prize competition's unusual "live typing" medium. What did this enable and disable, compared to the standard walkie-talkie, telegram-like, turn-taking style?

To make typing visible also, then, makes typing's "negative space" visible: hesitation. In a chat conversation where text is transmitted with every carriage return, only egregiously long pauses are taken to be "part" of the interaction. With more fluid and immediate feedback, silence acquires meaning. Failing to quickly answer a question in a face-to-face conversation, for instance, is tantamount in many cases to answering it. I recall asking a friend of mine how things were going with the girl he'd started seeing; the tiny pause and "um" in which he searched for the right words made it clear they were having problems. In other words, it wasn't his choice of words, but the choosing *itself*, that told the tale.[3]

I remember some Internet chat programs back in the '90s trying out this character-at-a-time approach, but people for the most part rejected it. It was too *invasive*, was the feeling: what people like about writing is the time and space to compose and edit before sharing it with the other person. The advantage of the character-at-a-time transmission, though, is that it starts to approach much more closely the condition of speech, with its fluidity of turn-taking and its choppy grammar, where what's lost in eloquence is made up for in agility.

Long missives weren't going to work as a strategy, as they have in certain years, where programs were able to steamroller the judges by eating up the clock and delivering ridiculously prolix answers. If two parties are taking strict turns under time pressure, they are putting themselves at the mercy of the length of the other's turns—here, for instance, is one program from 1996:

---

3. Apparently the world of depositions is changing as a result of the move from written transcripts to video. After being asked an uncomfortable question, one expert witness, I was told, rolled his eyes and glowered at the deposing attorney, then shifted uncomfortably in his chair for a full fifty-five seconds, before saying, smugly and with audible venom, "I don't recall." He had the transcript in mind. But when a *video* of that conversation was shown in court, he went down in flames.

```
REMOTE: How about telling me a joke?
JUDGE: No.
REMOTE: Here's my favorite joke... A dog limps into a bar with
    a bloody bandage on his foot. He's wearing a 10-gallon hat,
    six-shooters in his holsters and chaps. He growled, "I'm
    a-lookin' for the man that shot my paw."
```

"The joke's not funny . . . ," the judge writes, seemingly prompting the program to tell another one—which it does ("A knotty, worn-out old string walks into a bar . . ."). Meanwhile, three full minutes have elapsed.

Confederates, too, sometimes err on the epistolary side, as did a confederate in 1992, speaking on the topic of "Health":

```
JUDGE: well i generally eat various garbage that happens my
    way. to say i have a program would be really pushin it.
    Doritos are far and above my favorite feel good and groovy
    chow.
REMOTE: You have to be careful with garbage. That brings
    out another point. Natural health is about intuition to a
    great extent. When you eat too much garbage, you have to
    be careful of maintaining balance by your own intuition
    because sometimes, if you don't control it at this initial
    stage, you don't know until you see the results in your
    health that you have been harmed. High cholesterol is
    perfect example of this, and we all know its results;
    clogged arteries that are then very difficult to remedy, and
    there is controversy, over whether it is reversible.
```

This judge voted her a computer.

If a computer (or confederate) started rambling on too long in a "barge-in-able" test, the judge would just cut them off.

I realized something else about the character-by-character transmission, and what that might allow. Sometimes spoken dialogue becomes slightly nonlinear—as in, "I went to the store and bought milk and eggs, and on my way home I ran into Shelby—oh, and bread too," where we understand that bread goes with "bought" and not "ran into." (This is part of the function of "oh," another one of those words that traditional linguistics has had no truck with.) For the most part, though, there is so little lag time between the participants, and between the composition of a sentence in their minds and their speaking it out loud, that the subject matter rarely branches entirely into two parallel threads. In an instant-message conversation, the small window of time in which one person is typing, but the other cannot see what's being typed, is frequently enough to send the conversation in two directions at once:

```
A: how was your trip?
A: oh, and did you get to see the volcano?
B: good! how've things been back at the homestead?
A: oh, you know, the usual
B: yes we did get to see it!
```

Here the conversation starts to develop separate and parallel threads, such that each person's remark isn't necessarily about the most recent remark. It's possible that watching each other type eliminates the lag that creates this branching, although I had reason to believe it would do something else altogether . . .

Talking simultaneously for extended periods simply doesn't work, as our voice—emanating just inches away from our ears—mixes confusingly with our interlocutor's in the air and makes it hard to hear what they are saying. I was fascinated to learn that the deaf don't encounter this problem: they can comfortably sign while watching someone else sign. In large groups it still makes sense to have one "speaker" at a time, because people cannot look in more than one direction at a time, but conversations between *pairs* of signers, as Rochester Institute of Technology researcher Jonathan Schull observed, "involve more continuous simultaneous and overlapping signing among interlocutors" than spoken conversations. Signers, in other words, talk and listen at the same time. Schull and his collaborators conclude that turn-taking, even turn negotiation, far from being an essential and necessary property of communication, "is a reluctant accommodation to channel-contingent constraints."

One major difference between the Loebner protocols and traditional instant messaging is that, because the text is being created without any obvious ordering that would enable it to be arranged together on the screen, each user's typing appears in a separate area of the screen. Like sign language, this makes group conversation rather difficult, but offers fascinating possibilities for two-person exchange.

Another piece of my confederate strategy fell into place. I would treat the Turing test's strange and unfamiliar textual medium more like spoken and signed, and less like written, English. I would attempt to disrupt the turn-taking "wait and parse" pattern that computers understand and create a single, flowing duet of verbal behavior, emphasizing timing: whatever little computers understand about verbal "harmony," it still dwarfs what they understand about rhythm.

I would talk in a way that would, like a Ferneyhough piece, force satisficing over optimization. If nothing was happening on my screen, whether or not it was my turn, I'd elaborate a little on my answer, or add a parenthetical, or throw a question back at the judge—just as we offer and/or fill audible silence when we talk out loud. If the judge took too long considering his next question, I'd keep talking. I'm the one (unlike the bots) with something to prove. If I understood what the judges were writing, I'd spare them the keystrokes or seconds and jump in.

There's a trade-off, of course, between the number of opportunities for interaction and response, on the one hand, and the sophistication of the responses themselves. The former thrives with brevity, the latter with length. It seemed to me, though, that so much of the difficulty and nuance in conversation comes from understanding the question and offering an appropriate response—thus it makes sense in a Turing test to maximize the amount of interchanges.

Some judges, I would discover, would be startled or confused at this jumping of the gun, and I saw them pause, hesitate, yield, even start backspacing what they had half written. Other judges cottoned on immediately, and leaped right in after.[4]

In the first round of the 2009 contest, judge Shalom Lappin— computational linguist at King's College London—spoke with Cleverbot, and then myself. My strategy of verbosity was clearly in evidence: I made 1,089 keystrokes in five minutes (3.6 keystrokes a second) to Cleverbot's 356 (1.2/sec), and Lappin made 548 keystrokes (1.8/sec) in my conversation, compared to 397 (1.3/sec) with Cleverbot. Not only did I say three times as much as my silicon adversary, but I engaged the judge more, to the tune of 38 percent more typing from Lappin.

Looking back at the logs, though, I wanted to see if there was a way to quantify the *fluidity* of the human interactions against the rigidity of the machine ones. It occurred to me that you could create a benchmark—let's call it "swaps"—for the number of times that the party who typed the most recent keystroke changes.

I did manage to type three times as much as Cleverbot, but the *real* story, as it turns out, is in the swaps. Lappin's conversation with Cleverbot had 33 such events; his conversation with me had 492, almost fifteen times that.

There's no single obvious way to show, on the page, what that kind of interaction looks like (and this is partially the point, the gap between performance and transcript). His Cleverbot conversation, written in the traditional script style, looks a bit like this:

```
JUDGE: What language are you written in?
REMOTE: English.
JUDGE: Give me a break!
```

---

4. As Georgetown University linguist Deborah Tannen notes: "This all-together-now interaction-focused approach to conversation is more common throughout the world than our one-at-a-time information-focused approach."

```
REMOTE: Ok.
JUDGE: How do you enjoy the conference?
```

Even if we notate it more like musical staves, keystroke by keystroke,[5] there's very little more to the story. The keystrokes line up in tidy lines: statement/response/statement/response:

```
What_language_are_you_written_in?↵          Give_me_a_brea
                                    English.↵
```

```
k!↵     How_do_you_enjoy_the_conference?↵
  Ok.↵
```

Five carriage returns, four swaps.

His conversation with me, in traditional formatting, looks like this:

```
REMOTE: sweet, picking up an instrument
JUDGE: I meant Stones, Dylan, Beatles . . .
```

But the musical-staff-looking keystroke logs look utterly unlike the Cleverbot logs, and they tell a much different story:

```
I_ me  a nt_ S  t o n e s,_ D y lan   , _  B e  a
s we et,_  s tar ti n g_  a n «  ««« «« ««« « pi  ck
```

```
t  le s       . . .  ↵
  ing_ u p_ an_instru me n t↵
```

Two carriage returns, fifty-one swaps.

Alternately, we might try a third notation, which makes the difference even clearer: to string all the letters together in the order

---

5. We'll use "_" to mean a space, "↵" to mean carriage return/enter, and "«" to mean backspace.

typed, bolding the judge's keystrokes and leaving the computer's and my own unbolded. You get *this* from the human-computer dialogues:

```
What_language_are_you_written_in?↵English.↵Give_me_a_break!↵
Ok.↵How_do_you_enjoy_the_conference?↵
```

And *this* from the human-human dialogues:

```
sI_wemeet,a_nt_sStarttionnge_s,_aDay«lan«««,
««_«««B«epiackting_luep_san_instru.me.n.t↵↵
```

Now if that difference isn't night and day, I don't know what is. *Over.*

8. The World's Worst Deponent

Body (&) Language

Language is an odd thing. We hear communication experts telling us time and again about things like the "7-38-55 rule," first posited in 1971 by UCLA psychology professor Albert Mehrabian: 55 percent of what you convey when you speak comes from your body language, 38 percent from your tone of voice, and a paltry 7 percent from the words you choose.

Yet it's that 7 percent that can and will be held against you in a court of law: we are held, legally, to our diction much more than we are held to our tone or posture. These things may speak louder than words, but they are far harder to transcribe or record. Likewise, it's harder to defend against an accusation of using a certain word than it is to defend against an accusation of using a certain tone; also, it's much more permissible for an attorney quoting a piece of dialogue to superimpose her own body language and intonation—because they cannot be reproduced completely accurately in the first place—than to supply her own diction.

It's that same, mere 7 percent that is all you have to prove your humanity in a Turing test.

Lie Detection

One way to think about the Turing test is as a lie-detection test. Most of what the computer says—notably, what it says about itself—is false. In fact, depending on your philosophical bent, you might say that the software is *incapable* of expressing truth at all (in the sense that we usually insist that a liar must understand the meaning of his words for it to count as lying). I became interested, as a confederate, in examples where humans have to confront other humans in situations where one is attempting to obtain information that the other one doesn't want to give out, or one is attempting to prove that the other one is lying.

One of the major arenas in which these types of encounters and interactions play out is the legal world. In a deposition, for instance, most any question is fair game—the lawyer is, often, trying to be moderately sneaky or tricky, the deponent knows to expect this, and the lawyer knows to expect them expecting this, and so on. There are some great findings that an attorney can use to her advantage—for example, telling a story *backward* is almost impossible if the story is false. (Falsehood would not appear to be as modular and flexible as truth.) However, certain types of questions *are* considered "out of bounds," and the deponent's attorney can make what's called a "form objection."

There are several types of questions that can be objected to at a formal level. Leading questions, which suggest an answer ("You were at the park, weren't you?"), are out of bounds, as are argumentative questions ("How do you expect the jury to believe that?"), which challenge the witness without actually attempting to discover any particular facts or information. Other formally objectionable struc-tures include compound questions, ambiguous questions, questions assuming facts not yet established, speculative questions, questions that improperly characterize the person's earlier testimony, and cumulative or repetitive questions.

In the courtroom, verbal guile of this nature is off-limits, but it may be that we find this very borderline—between appropriate and inappropriate levels of verbal gamesmanship—is precisely the place where we want to position ourselves in a Turing test. The Turing test has no rules of protocol—anything is permissible, from obscenity to nonsense—and so interrogative approaches deemed too cognitively taxing or indirect or theatrical for the legal process may, in fact, be perfect for teasing apart human and machine responses.

Questions Deserving Mu

To take one example, asking a "simple" yes-or-no question might prompt an incorrect answer, which might provide evidence that the respondent is a computer. In 1995, a judge responded to "They have most everything on Star Trek" by asking, "Including [rock band] Nine Inch Nails?" The answer: an unqualified "Yes." "What episode was that?" says the judge. "I can't remember." This line of questioning goes some way toward establishing that the interlocutor is just answering at random (and is thereby probably a machine that simply doesn't understand the questions), but even so, it takes some digging to make sure that your conversant didn't simply misunderstand what you asked, isn't simply being sarcastic, etc.—all of which takes time.

Better might be throwing your interlocutor a loaded question, like the famous "Do you still beat your wife?" A question like this, asked, say, of a nonviolent unmarried heterosexual woman, is off at so many levels that it's basically unanswerable, requiring a huge backpedal and clarification of presumptions. Some languages actually have a term for answering questions like this, with the most iconic being the Japanese word *mu* that appears in certain Zen parables. "Does a dog have Buddha-nature, or does it not?" asks a student, and the master replies, *"Mu"*—meaning something like "All answers to this question are false." Or: "Your question is, itself, false." You can think of *mu* as a kind of "meta-no," an "unasking" of the question, or even as a kind of "runtime exception."[1] Lacking such a one-syllable recourse, though, a respondent is in the tight spot of needing to completely unpack and dismantle the question, rather than "responding" or "answering" as such. This is enough to fluster most humans, and flummox many,[2] and it's a good bet that a machine parser wouldn't have nearly the savvy to react appropriately.

Zero-Sumness

In looking at the way chess programs work, we discussed the "minimax" and "maximin" algorithm, two terms that we treated as synonymous. In "zero-sum" games, like chess, for one player to win necessitates that the other must lose—no "win-win" outcomes are possible—and so minimizing your opponent's outcome and maximizing your own constitute, mathematically anyway, the same strategy. (In the history of chess world champions, "prophylactic" players like Tigran Petrosian and Anatoly Karpov, playing for safety and to minimize their opponents' chances, stand alongside wild attackers like Mikhail Tal and Garry Kasparov, playing for chaos and to maximize their own chances.)

Here's one critical difference, perhaps the single biggest difference, philosophically, between conversation and chess. Asked by *Time* whether he sees chess "as a game of combat or a game of art," Magnus Carlsen, the current world number one, replies, "Combat. I am trying to beat the guy sitting across from me and trying to choose

---

1. Generally speaking, software has three ways of going awry: crashing while the code is being compiled into a program ("compile-time"), crashing when the program is being run by a user ("run-time"), or running smoothly but producing weird behavior. This is roughly analogous to sentences that are ungrammatical, un-meaningful, and false—to which we could reply "Huh!?," "*Mu*," and "Nope," respectively.
2. That Wikipedia contains relatively detailed instructions on how to parry such questions is indicative of how difficult they are to deal with.

the moves that are most unpleasant for him and his style. Of course some really beautiful games feel like they are art, but that's not my goal." Meaning, if there are collaborative elements, they are accidental by-products of the clash.

Capitalism presents an interesting gray space, where societal prosperity is more than the occasional by-product of fierce competition: it's the *point* of all that competition, from the society's point of view. Yet this non-zero-sum societal benefit is not something that any of the companies involved are *themselves* necessarily interested in, and it is not something that is, per se, guaranteed. (Ironically, we have antitrust laws that exist partially to *limit* the amount of collaboration between companies, as working *together* sometimes comes—e.g., price-fixing—at the consumer's detriment.) Whether you consider business zero-sum or non-zero-sum depends heavily on the context, and on your disposition.

But conversation, in the Turing test's sense of a "display of humanity," seems clearly and unambiguously non-zero-sum. Wit and repartee, for example, are chess's opposite: art that occasionally produces moments of what looks like sparring.

Seduction, interview, negotiation; you can read any number of books that portray these interactions in an adversarial light. E.g., interviewer Lawrence Grobel: "My job was to nail my opponent." In some cases—criminal trial being a good example—the adversarial mode may be unavoidable. But on the whole I think it's a mistake to consider conversations as zero-sum situations. Conversation at its best is less like minimax or maximin and more like "maximax." You put each other in a position to say great things. You play for the rally, not the score. You take pleasure in the alley-oop, the assist.

The Anti-Lincoln-Douglas

Of course the way the game is played depends in part on the way the game is scored: for instance, sports that celebrate, and tabulate, assists (ice hockey, for example, which gives credit to the last *two* players to touch the puck before the scorer) always seem to me to have more cohesion and team spirit among their players.

It breaks my heart, then, that so many of the communication "games" available to middle and high schoolers—namely, debate—feature conversation in its *adversarial,* zero-sum mode, where to weaken someone else's argument is as good as to strengthen your own. Additionally, the metaphors we use to describe dialectics, debate, and disagreement in our culture are almost exclusively *military*: *defending* a statement, *attacking* a position, *falling back* to a weaker version of a thesis, *countering* one accusation with another. But conversation is just as frequently a collaboration, an improvisation, a tangoing toward truth—not so much *duel* as *duet.* It'd be worth thinking about how to offer opportunities for our children to learn this, by reconsidering both our figurative speech and the extracurricular activities available to them.

Our legal system is adversarial, founded, like capitalism, on the idea that a bunch of people trying to tear each other apart, plus certain laws and procedures preventing things from getting too out of hand, will yield, in one, justice, and in the other, prosperity, for all. Sometimes this does happen; other times, it doesn't. At any rate, it's a terrible metaphor for the rest of life: I suppose we need Lincoln-Douglas debates and parliamentary debates and things like that in our high schools to train the lawyers of tomorrow, but how will we train the spouses and committee members and colleagues and teammates of tomorrow? We get to see how well presidential candidates can hack down, rebut, and debunk their rivals: How will we get to see how well they argue *constructively,* how they barter, coax, mollify, appease—which is what they will *actually* spend their term in office doing?

I propose the following: the Anti-Lincoln-Douglas, Anti-parliamentary debate. Two sides are given a set of distinct and not obviously compatible objectives: one team, for instance, might be striving to maximize individual liberty, and the other might be striving to maximize individual safety. They are then asked to collaborate, within strict time limits, on a piece of legislation: say, a five-point gun-control bill. After the exact language of the bill is drafted, each team will independently argue to a judging panel why the legislation supports their side's goal (liberty on the one, safety on the other), and the judges will award a score based on how convincingly they make that case.

The tournament judges then give both sides the *same* score, which would be the *sum* of those two scores.

It's that simple. You pair up with each team in the tournament, round-robin-style, and the team with the most points at the end wins. No individual *match* has a victor, yet the tournament as a *whole* does, and they became so by working *with* each team they were paired with. The point structure encourages both groups to find mutually agreeable language for the bill—or else they will have nothing to present the judges at all—and, even beyond that, to help each other "sell" the bill to their respective constituencies.

Imagine the national Lincoln-Douglas champion and the national Anti-Lincoln-Douglas champion: Which one would you rather attend a diplomatic summit? Which one would you rather be married to?

Jewel-Tone Rubber Blobs

Success in distinguishing when a person is lying
and when a person is telling the truth is highest
when . . . the interviewer knows how to encour-
age the interviewee to tell his or her story.

—PAUL EKMAN, *TELLING LIES*

Put practically and more generally, a collaborative, "maximax" style of conversation means that you speak with a mindfulness toward what the other person might be able to say next. As far as answers to "How are you?" go, "Good" is probably down there with the worst. "Good, you?" or "Good, what's up?" don't give much of an answer, but transfer the momentum back to the asker without much friction. "Ugh . . ."

and to a lesser extent "Amazing!" invite inquiry, and this effect is increased by alluding, however vaguely, to recent events: "Yesterday sucked; today was awesome" or "Not so good *today* . . ." or "Better!" or even the subtle "Good, actually," whose "actually" hints at some reason to expect otherwise. It's terse, but intriguing enough to work.

Think of these elements, these invitations to reply or inquiry or anecdote, topic switch, exposition, you name it, as akin to an indoor rock-climbing gym's "holds"—those bright jewel-tone rubber blobs that speckle the fake rock walls. Each is both an aid to the climber and an invitation onto a certain path or route along the ascent.

This notion of holds explains and synthesizes all sorts of advice about conversation. For instance, business networking experts and dating/seduction gurus alike recommend wearing one at least slightly unusual item of clothing or accessory. In *How to Talk to Anyone,* Leil Lowndes calls these items "Whatzits," and in *The Game,* Mystery and Neil Strauss dub the practice "peacocking." The principle's the same: you give other people an easy first hold—a simple and obvious way to initiate a conversation with you, if they want. The other day I ran into a friend of a friend at an art gallery opening, and wanted to strike up a conversation but wasn't sure how. All of a sudden I noticed he was wearing a vest—a rarity—and so my first remark became obvious— "Hey, nice vest!"—and once the conversation was in motion, it was easy to keep it going from there. It's interesting to consider: dressing generically might actually be a kind of *defense,* presenting a rock face with no holds, making yourself harder to chat up. All clothing can be armor.

The Mystery/Strauss camp find themselves *opposed* to the conventional wisdom of folks like Lowndes, Larry King, and Dale Carnegie in one particular, however: they don't advise the asking of questions. Instead of asking whether someone has any siblings, they counsel us to say, "You seem like an only child to me." There are a couple reasons for this approach: some bogus, some legitimate.

The bogus reason is that it comes off as less interested in the person than asking them directly. The Mystery/Strauss camp are obsessed with *status* in conversation—which is a game that Larry King, for example, or Charlie Rose doesn't have to play: the interviewer's *job* is to be interested in the other person. Cool people, Mystery and Strauss seem to be saying, are more interested in holding forth than in learning about other people. To be fair, let's consider the context: these guys are talking about picking up supermodels and celebrities in L.A.—so maybe games of status matter more in those circles. (I've heard, for instance, that one of the *best* openers for normal folks is in fact just about the worst and most insulting thing you can say to a Hollywood actor: "So, what have you been up to lately?") As for me, I celebrate the sexiness of enthusiasm. And I submit that the *truly* cool people don't *care* that they seem interested.[3] The only thing sexier than curiosity is confidence, and the person with both will simply ask away.

Besides, the kind of guardedness that comes from developing an entire "method" to the way you talk to people suggests a kind of minimax approach to wooing—by avoiding umpteen pitfalls, you can indeed minimize rejections, but that's playing not to lose, to maximize the *minimum* outcome.[4] Whereas authenticity and genuineness, which maximize the *maximum* outcome, succeed perhaps less often but more spectacularly.

A legitimate reason to prefer, if you do, statements to questions is that the statement (e.g., "You seem like an only child") simultaneously asks the question and hazards a guess. The guess is intriguing—we *love* to know what other people think of us, let's be honest—and so now we have at least two distinct holds as far as our reply: to answer the question and to investigate the reason for the guess.

The disadvantage to questions is that you leave too few holds for

---

3. Also, there's no point in trying to mask your interest—whether it be sexual, social, academic, professional, or other—in the person anyway, because the very fact that you're talking to them signals it: they're not stupid.
4. A common complaint among "pickup artists," I learned, is that they get tons of phone numbers but no one calling back—a telltale sign of a maximin approach.

the person to learn about *you*—but statements don't really go too far in this direction either. A tiny anecdote that springs a question works, I think, best. The other person can either ask *you* about your anecdote or answer the question. Whatever they want.

Another manifestation of the climbing hold is the hyperlink. The reason people can lose themselves in Wikipedia for hours is the same reason they can lose themselves in conversation for hours: one segue leads to the next and the next. I sometimes get a kind of manic, overwhelmed sensation from conversation when there seem to be almost *too* many threads leading out of the page.[5] These are the "Aah, where do I even begin!" moments. It's not necessarily a pleasant feeling, but it's much more pleasant than the opposite, the cul-de-sac, sink vertex, sheer cliff, the "Now what?," the "So . . ."

This feeling is frustrating, stultifying, stymieing, but also kind of *eerie*—eerie in the way that a choose-your-own-adventure page with no choices at the bottom is eerie. You get to the end of the paragraph: What the hell? What next?

I went on a date in college with the assistant stage manager of a play I was sound designing—we'd hit it off talking about Schopenhauer, I think it was, one day after rehearsal. So when I met her at her building on a Sunday afternoon and we walked over to catch the matinee of the other play running that weekend on campus, I started from the two holds I had: "So, what kinds of stuff do you like to do when you're not, you know, running light boards or thinking about German philosophers?" Inexplicably, she testily reproached me: "*I* don't *know*!" And I waited for the rest of her answer—because people often do that, say "*I* don't *know* . . ." and then say something—but that was it, that was the whole answer.

---

5. Graph theory talks about the "branching factor" or the "degree" of a vertex, meaning the number of nodes in the graph to which a given node connects. The conversational analogue is how many distinct continuations or segues there are from the present remark or topic; for my money, the sweet spot is around two or three.

Your mileage may vary: such extreme examples of someone deliberately jamming the conversational cogs are pretty rare. But sometimes we present a sheer face to someone by accident, and with the best of intentions. Leil Lowndes talks about meeting a woman who was hosting an event where Lowndes was speaking, and the woman just sits there and waits for the "conversation expert" to dazzle her. The momentum already slipping out of the interaction, Lowndes tries to keep it going by asking the host where she's from. "Columbus, Ohio," the host says, and then just smiles expectantly, to see what the pro will say next. But where can anyone—who doesn't happen to have anecdotal experience of Columbus, Ohio—go from there? The only avenue is to offer your own background uninvited ("Ah, well I'm from _____") or just say something like "Oh, I don't know much about Columbus, although you hear about it a lot; what's it like?" Either way, the holds are far from obvious.

The same idea applies in the case of text-based role-playing games, a.k.a. "interactive fictions," some of the earliest computer games. 1980's *Zork*, for instance, perhaps the best-known (and best-selling) title of the genre and era, begins as follows: "You are standing in an open field west of a white house, with a boarded front door. There is a small mailbox here." That's just about right. Two or three different holds, and you let the user choose.

My friend and I once jokingly tried to imagine the world's most unhelpful text-based role-playing game. "You are standing there," it would begin. Or, if you enter a house: "You enter the house." And if the user types the "look" command: "You see the inside of the house." No description at all. Sheer walls, literally and figuratively.

There's only one or two exceptions: you might purposely try to strip the holds off a particular story because you want *less* participation from your conversant. I often find myself saying things like "So I rode my bike to the coffee shop this afternoon, and there was this *guy* there, and he was all—" "You *rode* your *bike*? In *this* weather?" And the wind goes out of the sails. Sometimes you want your listeners to choose their own adventure; sometimes you want to take them on one adventure in particular. Replacing "rode my bike" with "went," reducing the degree of that vertex, eliminates a potentially distracting hold. It reduces conversational drag. And maybe forcing my listener to envision the bicycle is a waste of their brainpower, a red herring anyway.

The bottom line is that it depends on what our conversational *goals* are. If I'm just shooting the breeze, I'll put in all kinds of holds, just to see what my conversant grabs onto—just to give them the most options for a favorable topic direction. If I have something really important to say, I'll streamline.

When you want to artfully wrap up a conversation, it's easy to put on the brakes. You stop grabbing their holds, you stop free-associating ("that reminds me of . . ."), you start stripping the holds off of your own turns. Eventually the conversation downshifts or cul-de-sacs and you end it. It's subtle and graceful, sometimes even subliminal.

Specificity: Infie-J

A friend from across the country calls me just to catch up. "What are you up to?" she says.

Where I might have said, before my Turing test preparation, "Oh, nothing," or "Oh, just reading," now I know to say *what* I'm reading, and/or what I'm reading *about*. Worst case, I waste, you know, a dozen or so syllables of her time. But even then, I'm displaying a kind of enthusiasm, a kind of zest, not only with respect to my own life, but with respect to the conversation as well: I'm presenting *an uneven face*. In the sense of: climbable. I'm offering departure points. This is why "peacocking" makes sense; this is why it's good to decorate your house with photographs from your life, especially of travel, and with favorite books. A good house, from the perspective of both conversation and memory, is neither squalid (full of meaningless things) nor sterile (devoid of anything), but abounds in (metaphorical) jewel-tone rubber blobs.

So, making the most minor of adjustments, I say, "Reading *Infinite*

Jest," and she says, "Oh! *Infie-J!*" and I say, "You call it *Infie-J*?!" then we are off to the races before I even have a chance to ask her how *she* is—and whenever the *Infie-J* thread runs out of steam, I will—and meanwhile we've set a precedent that we don't want the short, polished, seamless answer. It's the seams on a baseball, for instance, that allow it to curve.

Game Time

All the theory is well and good, but what about the practice? How to bring the idea of holds to the Turing test?

Holds are useful for the judges to manipulate. Limiting the number of holds can stall the conversation, which is potentially interesting: humans, on the side of truth, would have more of an incentive to reanimate it than the computers. On the other hand, computers are frequently oblivious to conversational momentum anyway, and tend to be eager to topic shift on a dime; it's likely a many-holds approach from the judge might be best. One idea that a judge might employ is to put something odd into the sentence—for instance, if asked how long they traveled to the test: "Oh, just two hours in the ol' Model T, not too far." A grammatical parser might strip the sentence down to "two hours—not far," but a human will be so intrigued by the idea of a guy driving a hundred-year-old car that the ordinary follow-ups about traffic, commuting, etc., will be instantly discarded.

As for myself on the confederate side, in that odd and almost oxymoronic situation of high-pressure chitchat, I would have planted holds all over the first few remarks ("no holds barred"?), because there just isn't any time for slow starts. A judge might find it useful to stall a confederate, but surely not the reverse.

A simple, flat, factual answer (what Lowndes calls a "frozen" or "naked" answer) offers essentially a single hold, asking for more information about that answer. (Or one and a half holds, if you count awkwardly volunteering one's *own* answer to the same question:

"Cool, well *my* favorite movie is . . .") The only thing worse—which many bots and some confederates nonetheless do—is not answering at all. Demurrals, evasions, and dodges in a Turing test can be fatal: it's harder to prove that you *understand* a question when you're dodging it.

It surprised me to see some of the other confederates being coy with their judges. Asked what kind of engineer he is, Dave, to my left, answers, "A good one. :)" and Doug, to my right, responds to a question about what brings him to Brighton with "if I tell you, you'll know immediately that I'm human ;-)." For my money, wit is very successful, but coyness is a double-edged sword. You show a sense of humor, but you jam the cogs of the conversation. Probably the most dangerous thing a confederate can do in a Turing test is *stall.* It's suspect—as the *guilty* party would tend to be the one running out the clock—and it squanders your most precious resource: time.

The problem with the two jokes above is that they are not contextually tied in to anything that came before in the conversation, or anything about the judges and confederates themselves. You could theoretically use "if I tell you, you'll know immediately that I'm human" as a wild-card, catch-all, panic-button-type answer in a bot (similar to ELIZA's "Can you say more about that?"), applicable for virtually *any* question in a conversation. And likewise, it's easy to imagine a bot replying "A good one :)" by template-matching a question asking what kind or type of *x* something is. Decontextual, non-context-sensitive, or non-site-specific remarks are, in the case of the Turing test, dangerous.

Answer Only the Question Asked

Many last names in America are "occupational"—they reflect the professions of our ancestors. "Fletchers" made arrows, "Coopers" made barrels, "Sawyers" cut wood, and so on. Sometimes the alignment of one's last name and one's career is purely coincidental—see, for instance, poker champion Chris Moneymaker,[6] world record-holding sprinter Usain Bolt, and the British neurology duo, who sometimes published together, of Russell Brain and Henry Head. Such serendipitous surnames are called "aptronyms," a favorite word of mine.

Such were the sorts of thoughts in my head when I called attorney Melissa Prober. Prober's worked on a number of high-profile cases, including being part of the team that defended President Clinton during the investigation leading to his impeachment hearings and subsequent Senate acquittal. The classic advice given to all deponents, Prober explained to me, is to answer *just* the question being asked, and *only* the question being asked.

Her colleague (who has since become the executive assistant U.S. attorney for the district of New Jersey) Mike Martinez concurred. "If you volunteer too much— First, it's just not the way the system's supposed to be anyway. The way it's supposed to be is Lawyer A makes a question and Lawyer B decides whether that's a fair question. If the person answers beyond that, then he's doing so unprotected."

It's interesting—many Loebner Prize judges approach the Turing test as a kind of interrogation or deposition or cross-examination; strangely, there are also a number of *confederates* who seem to approach it with that role in mind. One of the conversations in 2008 seems never to manage to get out of that stiff question-and-response mode:

```
JUDGE: Do you have a long drive?
REMOTE: fairly long
JUDGE: so do I :( ah well, do you think you could have used
    public transport?
REMOTE: i could have
JUDGE: and why not?
REMOTE: i chose not to
```

---

6. Apparently his German ancestors, surname Nurmacher, were in fact "moneyers," or coin smiths, by trade.

```
JUDGE: that's fair. Do you think we have too many cars on the
    road or not enough today?
REMOTE: its not for me to say
```

Yawn! Meanwhile the computer in the other terminal is playful from the get-go:

```
JUDGE: HI
REMOTE: Amen to that.
JUDGE: quite the evangelist
REMOTE: Our Father, who art in cyberspace, give us today our
    daily bandwidth.
JUDGE: evangelist / nerd lol. So how are things with you today?
```

And has practically sealed up the judge's confidence from sentence two. Note that the confederate's stiff answers prompt more grilling and forced conversation—what's your opinion on such-and-such political topic? But with the computer, misled into assuming it's the real person by its opening wisecracks, the judge is utterly casual: How are things? This makes things easier for the computer and harder for the confederate.

On the Record

The humans in a Turing test are strangers, limited to a medium that is slow and has no vocal tonality, and without much time—and also stacked against them is the fact that the Turing test is *on the record*.

In 1995 one of the judges, convinced—correctly, it turns out—that he was talking to a female confederate, *asked her out,* to which she gave the *mu*-like non-answer "Hm. This conversation is public isn't it?" And in 2008 two humans got awkward and self-conscious:

```
JUDGE: Did you realise everyone can see what's being typed on
    this machine on a big screen behind me?
```

REMOTE: uhh.. no.

REMOTE: so you have a projector hooked up to your terminal then?

JUDGE: Yeah, it's quite freaky. So watch what you say!!

That guardedness makes the bots'—I can't believe I was going to say "*lives*" here—easier.

As author and interviewer David Sheff—who wrote, among numerous books and articles, the last major interview with John Lennon and Yoko Ono, for *Playboy* in 1980—explains to me, "The goal has always been to transform the conversation from a one that is, you know, perceived by the subject *as* an interview to one that becomes a dialogue between two people. The best stuff always came when it was almost as though the microphone disappeared." In the conversation we saw earlier, with the judge saying "Do you think we have too many cars on the road" in one window and "So how are things with you today?" in the other, this difference in timbre can make a big difference.

The paradigm of guardedness in our culture is the politician. Just the other day some of my friends were talking about a mutual acquaintance who has started obsessively scrubbing and guarding his Facebook profile. "What, is he running for office or something?" they half joked. That's the kind of character-sterility that our society both demands and laments in politics. No holds.

Professional interviewers across the board say that guardedness is the worst thing they can run into, and they are all, as far as I can tell, *completely* unanimous in saying that politicians are the worst interview subjects imaginable. "With every response they're [politicians] trying to imagine all the pitfalls and all the ways it could come back to bite them," Sheff says. "The most interesting people to interview are the people who want to do exactly what you want to do in this test—which is to show that they're a unique individual." That tends not to be on politicians' agendas—they treat conversation as a minimax game, partially because their worst words and biggest gaffes and failures so often ring out the loudest: in the press, and sometimes also in history. Whereas *artists,* for example, will generally be remem-

bered for their *best,* while their lesser works and miscues are grace-
fully forgotten. They can be non-zero-sum.

Prolixity

The more words spoken the better the chance of
distinguishing lies from truthfulness.

—PAUL EKMAN

Add to all the above the fact that the Turing test is, at the end of
the day, a race against the clock. A five-second Turing test would
be an obvious win for the machines: the judges, barely able even to
say "hello," simply wouldn't be able to get enough data from their
respondents to make any kind of judgment. A five-hour one would
be an obvious win for the humans. The time limit at the Loebner
Prize contest has fluctuated since its inception, but in recent years
has settled on Turing's original prescription of five minutes: around
the point where conversation starts to get interesting.

Part of what I needed to do was simply to make as much engage-
ment happen in those minutes as I physically and mentally could.
Against the terseness of the deponent I offered the prolixity and log-
orrhea of the author. In other words, I talked a *lot.* I only stopped
typing when to keep going would have seemed blatantly impolite or
blatantly suspicious. The rest of the time, my fingers were moving.

If you look at Dave's transcripts, he warms up later on, but starts
off like he's on the receiving end of a deposition, answering in a kind
of minimal staccato:

```
JUDGE: Are you from Brighton?
REMOTE: No, from the US
JUDGE: What are you doing in Brighton?
REMOTE: On business
JUDGE: How did you get involved with the competition?
REMOTE: I answered an e-mail.
```

Like a *good* deponent, he lets the questioner do all the work[7]—whereas I went out of my way to violate that maxim of "A bore is a man who, being asked 'How are you?' starts telling you how he is." (And I might add: "And doesn't stop until you cut him off.")

```
JUDGE:  Hi, how's things?
REMOTE: hey there
REMOTE: things are good
REMOTE: a lot of waiting, but...
REMOTE: good to be back now and going along
REMOTE: how are you?
```

When I saw how stiff Dave was being, I confess I felt a certain confidence—I, in my role as the world's worst deponent, was perhaps in fairly good shape as far as the Most Human Human award was concerned.

This confidence lasted approximately sixty seconds, or enough time to glance to my other side and see what Doug and *his* judge had been saying.

Fluency

Success in distinguishing when a person is lying and when a person is telling the truth is highest when . . . the interviewer and interviewee come from the same cultural background and speak the same language.

—PAUL EKMAN

In 2008, London *Times* reporter Will Pavia misjudged a human as a computer (and thus voted the computer in the other window a

---

7. Prober recalled asking one deponent if he could state his name for the record. His answer: "Yes."

human) when a confederate responded "Sorry don't know her" to a question about Sarah Palin—to which he incredulously replied, "How can you possibly not know her? What have you been doing for the last two months?" Another judge that year opened his conversations with a question about the "Turner Prize shortlist," the annual award to a contemporary British visual artist, with similarly hit-or-miss results: Most Human Computer winner Elbot didn't seem to engage the question—

```
JUDGE: What do you think of this year's Turner Prize
    shortlist?
REMOTE: Difficult question. I will have to work on that and
    get back to you tomorrow.
```

—but neither, really, did the confederate in that round:

```
JUDGE: What do you think of this year's Turner Prize shortlist?
REMOTE: good I think. Better than the years before i herad
JUDGE: Which was your favorite?
REMOTE: Not really sure
```

Runner-up for 2008's Most Human Computer was the chatbot "Eugene Goostman," which pretended to be an *immigrant,* a non-native speaker of English with an occasionally shaky command of the language:

```
REMOTE: I am from Ukraine, from the city called Odessa. You
    might have heard about it.
JUDGE: cool
REMOTE: Agree :-) Maybe, let's talk about something else? What
    would you like to discuss?
JUDGE: hmm, have you heard of a game called Second Life?
REMOTE: No, I've never heard such a trash! Could you tell me
    what are you? I mean your profession.
```

Is this cheating, or merely clever? Certainly it's true that if language is the judge's sole means of determining which of his correspondents is which, then any limitations in language use become limitations in the judge's overall ability to conduct the test. There's a joke that goes around in AI circles about a program that models catatonic patients, and—by saying nothing—perfectly imitates them in the Turing test. What the joke illustrates, though, is that seemingly the less fluency between the parties, the less successful the Turing test will be.

What, exactly, does "fluency" mean, though? Certainly, to put a human who only speaks Russian in a Turing test with all English speakers would be against the spirit of the test. What about dialects, though? What exactly counts as a "language"? Is a Turing test peopled by English speakers from around the globe easier on the computers than one peopled by English speakers raised in the same country? Ought we to consider, beyond national differences, demographic ones? And where—as I imagine faltering against a British judge's cricket slang—do we draw the line between *language* and *culture*?

It all gets a bit murky, and because in the Turing test all routes to and from intelligence pass through language, these become critical questions.

All of a sudden I recalled a comment that Dave Ackley had made to me on the phone, seemingly offhand. "I really have no idea how I would do as a confederate," he said. "It's a little bit of a crapshoot whether the judges are your kind of people." He's right: if language is the medium with which we confederates must prove ourselves to the judges, then there are any number of things that can aid or impair it, from shared interests or reference points, to generational gaps, to nuances of allusion and slang.

Among the four confederates, Dave and I are Americans, Doug is Canadian, and Olga is a Russian-born South African. Among the four judges, two are English, one is an American expatriate to England, and one is a Canadian. I had read logs of Loebner Prizes past and had seen the problems that arise when cultural mismatch or cultural disfluency rears its head.

I wondered: Would any such cultural issues come to bear in 2009? All my preparations, my investigations, all the good advice I'd gotten from lawyers, linguists, researchers, and interviewers, wilted compared to actually having something in common and hitting it off with someone. To "speaking the same language," however literally or figuratively. Would that play in this year?

I didn't have to wait long for my answer; any uncertainty I'd had on that score, not to mention the optimism I'd begun to feel about my own chances, faded fast when I glanced at Doug's terminal:

```
JUDGE: Hey Bro, I'm from TO.
REMOTE: cool
REMOTE: leafs suck
REMOTE: ;-)
JUDGE: I am jist back froma sabbatical in the CS Dept. at U
  or T.
REMOTE: nice!
JUDGE: I remember when they were a great team.
JUDGE: That carbon date me, eh?
REMOTE: well, the habs were a great team once, too...
REMOTE: *sigh*
JUDGE: YEH, THEY SUCK TOO.
REMOTE: (I'm from Montreal, if you didn't guess)
```

Doug and his judge had just discovered that they were both from Canada. And they started to let rip with abbreviations and nicknames and slang and local references. *And they started to talk about hockey.*

I was in trouble.

9. Not Staying Intact

Each is trying not to give himself or herself away,
each is preserving fundamental loneliness, each
remains intact and therefore unfructified. In such
experiences there is no fundamental value.

—BERTRAND RUSSELL

A tree or shrub can grow and bloom. I am always
the same. But I am clever.

—RACTER

Meet Converse

As I read through the Loebner Prize transcripts, reams (literally) of small talk and chitchat and "how's it going?" and basic world-knowledge questions, I was struck by the following dialogue from the 1997 contest:

```
REMOTE: Hi, my name is Catherine, what's yours?
JUDGE: My name is Elliott.
REMOTE: Well, it's nice to meet you, Elliott. Did you see that
    story on CNN last night about the lesbian couple who came
    out at a White House party on Sunday?
JUDGE: Yes, I did. I think it may all be a publicity stunt for Ellen.
```

REMOTE: Ellen Degeneres was one of them - she was kissing her lover in front of Bill Clinton. Do you think he was very embarrassed?

JUDGE: I think he want the gay vote. He will do anything to get it and keep it. He does not have a conservative base.

REMOTE: Are you still a Friend of Bill Clinton?

JUDGE: I am not sure if he has any real friends. He has achieved alot and has left alot of people behind.

REMOTE: You really don't like the guy--well, at least that's definite--I thought most people were just indifferent or numb by now.

JUDGE: I think we could have a more honest and dynamic leader. One able to face the real dangers facing the country and address them with courage and intelligence. Rather then listening to public opinion polls.

REMOTE: I guess that means you really think there's serious behind Whitewater, the Special prosecutor and all that stuff?

JUDGE: We have two dead people, Brown and guy who commited suicide, we have lost papers. And we have Clintons people willing to tell all. I think that Hillary will be found guilty of something.

REMOTE: But how do you actually feel about him, now in the number two term?

JUDGE: He is too smart to get caught. He deserved to named hotel manager of the year.

REMOTE: OK, but what happens now--does it all drift on, with everyone waiting to see how much his friends have said about him under threat of lots and lots of jail--or will something dramatic actually happen?

I have to say, this is up there as one of the most trenchant, coherent, focused, and substantive conversations in the entire annals of the Turing test.

Which is why it's so surprising that Catherine is a computer.

Specifically, she's a program called Converse, written by science author, entrepreneur, and International Computer Games Association president David Levy, along with University of Sheffield AI professor (now emeritus) Yorick Wilks. This conversation really shook me—how did they do that? It shook me, that is, until I read the *other* judges' conversations with "Catherine." Observe:

> REMOTE: Are you still a Friend of Bill Clinton?
>
> JUDGE: Never met him.
>
> REMOTE: You really don't like the guy--well, at least that's definite--I thought most people were just indifferent or numb by now.
>
> JUDGE: Have you seen the movie Mars attacs?
>
> REMOTE: Give me a break -- I'm just here to answer dumb questions -- not that! What were we talking about... Ok.. I guess that means you really think there's serious behind Whitewater, the Special prosecutor and all that stuff?
>
> JUDGE: I'm cold.
>
> REMOTE: But how do you actually feel about him, now in the number two term?
>
> JUDGE: Hey, don't you get it? I don't want to talk about Clinton.
>
> REMOTE: Sorry -- my attention slipped for a moment--can we pushon to something more gripping. where was I.. oh yes OK, but what happens now--does it all drift on, [. . .]

Amazingly, the program only has a single conversational path—a nicely site-specific one, having been written the night before the contest about a current event—and all of its text-parsing and programming finesse is aimed at keeping the conversation on it. The feeling of eeriness that I'd felt reading that first conversation disappeared; it was actually quite simple indeed, and when it crashed and burned, it did so rather spectacularly. But when it worked, it *really* worked.

Another judge—Clay—gets taken for the same ride as Elliott, waxing political, musing over the state of the nation, and really engaging the questions. In fact everything is going swimmingly until the very end, when Clay's signing off:

```
JUDGE: it looks like i've got to go, Catherine. its been nice
    talking with you
JUDGE: do you need water or something before the 5 minute
    rounds?
```

At which point the program erupts into gibberish.

```
REMOTE: OK, yes on balance... Time to get off this one I think
    and onto something more down to earth!
JUDGE: like what?
REMOTE: Almost certainly! let's move on
REMOTE: I suppose it depends on where you're coming from, but
    as the song goes in My Fair Lady, "Why can't a woman be
    more like a man?"
```

Defeat from the jaws of victory.

The whole thing got me thinking about conversation, and storytelling in particular. There was a program called "Joe the Janitor" in 1995—written by 1994-winner Thomas Whalen of the Canadian Communications Research Centre—that took a similar tack, adopting the persona of a janitor in danger of being fired after being wrongly accused of stealing from his office. The template matching was sophisticated enough that references to "unions" would prompt him to gripe about his union and so on—a level of responsiveness above Catherine—and the "script" itself was quite engaging and sympathy inducing. Joe's depressed and obsessing demeanor—reasonable, given his situation—helped make his topic monomania seem plausible:

```
JUDGE: ever bneen to austraila?
REMOTE: Huh?
REMOTE: Maybe I just got too much on my mind.
REMOTE: I got real trouble at work. I think I'm going to get
   fired.
```

And, like many depressed people, he didn't take the judges' suggestions for ways to help the situation, or attempts to lighten his spirits, to heart. He just kept depressively spieling.

Hot-Blooded and Cold-Blooded

The second chatbot to achieve major prominence within the AI literature, after ELIZA in 1966, was a bot written in 1972 called PARRY. In many ways, PARRY is ELIZA's opposite: where ELIZA mimicked a therapist, PARRY mimicked a patient—specifically, a paranoid schizophrenic. Where ELIZA was satirical, a "parody" of a therapist (in Weizenbaum's words), PARRY—written by Kenneth Colby, a Stanford psychiatrist—was an earnest attempt to understand and model paranoia.

Stylistically, PARRY is also ELIZA's inverse. While ELIZA practiced the "reflective listening" of Carl Rogers and his school of "client-centered therapy," PARRY (after which Catherine was modeled) took the lead—and kept it. When these two programs, in the early 1970s, were made to talk with each other, these differences were clear as day. Note how the attempt to change topic is unceremoniously rebuffed:

```
PARRY: I went to the track at Bay Meadows a while back.
ELIZA: What does that suggest to you?
PARRY: I went to the races.
ELIZA: What else would you like to discuss?
PARRY: Horseracing interests me.
```

In this conversation, and in general, ELIZA seems only to view itself as a kind of verbal mirror, an opportunity for its interlocutor to speak. PARRY seems only to view its interlocutor as an opportunity for *it* to speak. If conversation were tennis, the former is a practice wall, and the latter, a serving machine. Neither really needs to engage or react—or even move.

As such, they perfectly illustrate the Scylla and Charybdis of conversation: at one extreme is ELIZA, what you could call the "reptilian" or "cold-blooded" approach—"not enough me." At the other is PARRY, "too much me," what could be called the "hot-blooded" or "mammalian" approach. As cognitive neuroscientist Michael Gazzaniga writes, "Vocal communication from the chimp's point of view may just be 'It's all about me,' which when you think about it isn't all that different from many human dates."

Speaking of dates: Arguably the two most prominent "pickup artists" of the past twenty years, Mystery and Ross Jeffries, fall into the same dichotomy. Mystery, star of *The Game* as well as VH1's *The Pickup Artist*, was a magician in his twenties; he first learned the gift of gab as *patter*: a way to hold and direct a person's attention while you run them through a routine. "Looking back on the women I have shared intimate moments with," he writes, "I just talked their ears off on the path from meet to sex . . . I don't talk about her. I don't ask many questions. I don't really expect her to have to say much at all. If she wants to join in, great, but otherwise, who cares? This is my world, and she is in it." This is the performer's relationship to his audience.

At the other extreme is the therapist's relationship to his client. Ross Jeffries, arguably the most famous guru of attraction before Mystery,[1] draws his inspiration not from stage magic but from the same field that inspired ELIZA: therapy. Where Mystery speaks mostly in the first person, Jeffries speaks mostly in the second. "I'm gonna tell you something about yourself," he begins a conversation with one woman. "You make imagery in your mind, very, very vividly; you're a very vivid

---

1. And allegedly the inspiration for Tom Cruise's character in *Magnolia* (for which Cruise received an Oscar nomination and a Golden Globe).

daydreamer." Where Mystery seems perhaps a victim of solipsism, Jeffries seems almost to *induce* it in others.

Jeffries's approach to language comes from a controversial psychotherapeutic and linguistic system developed in the 1970s by Richard Bandler and John Grinder called Neuro-Linguistic Programming (NLP). There's an interesting and odd passage in one of the earliest NLP books where Bandler and Grinder speak disparagingly of talking about oneself. A woman speaks up at one of their seminars and says, "If I'm talking to someone about something that I'm feeling and thinking is important to me, then . . ."

"I don't think that will produce connectedness with another human being," they respond. "Because if you do that you're not paying attention to *them,* you're only paying attention to *yourself.*" I suppose they have a point, although connectedness works both ways, and so introspection could still connect *them* to *us,* if not vice versa. Moreover, language at its best requires both the speaker's motive for speaking *and* their consideration of their audience. Ideally, the other is in our minds even when we talk about ourselves.

The woman responds, "OK. I can see how that would work in therapy, being a therapist. But in an intimate relationship," she says, it doesn't quite work. I think that's true. The therapist—in some schools of therapy, anyway—wants to remain a cipher. Maybe the interviewer does too. *Rolling Stone* interviewer Will Dana, in *The Art of the Interview,* advises: "You want to be as much of a blank screen as possible." David Sheff remarked to me that perhaps "the reason I did so many interviews is that it was always more comfortable to talk about other people more than talk about myself."[2] In an interview situation, there's not necessarily anything wrong with being a blank screen. But a *friend* who absents himself from the friendship is a bit of a jerk. And

---

2. At the same time, he says, he attributes some of his success as an interviewer to a rapport that came from "an openness about myself—it was my nature to talk about whatever I was going through, in a way that wasn't meant to disarm but it did disarm."

a lover who wants to remain a cipher is sketchy in both senses of the word: roughly outlined, and iffy.

The Limits of Demonstration

If poetry represents the most *expressive* way of using a language, it might also, arguably, represent the most *human*. Indeed, there's a sense in which a computer poet would be much scarier of a prospect to contend with than a computer IRS auditor[3] or a computer chess player. It's easy to imagine, then, the mixture of skepticism, intrigue, and general discomfort that attended the publication, in 1984, of the poetry volume *The Policeman's Beard Is Half Constructed*: "The First Book Ever Written by a Computer"—in this case, a program called Racter.

But as both a poet and a programmer, I knew to trust my instincts when I read *The Policeman's Beard Is Half Constructed* and felt instantly that something was fishy.

I'm not the only one who had this reaction to the book; you still hear murmurs and grumbles, twenty-five years after its publication, in both the literary and AI communities alike. To this day it isn't fully known how exactly the book was composed. Racter itself, or some watered-down version thereof, was made available for sale in the 1980s, but the consensus from people who have played around with it is that it's far from clear how it could have made *The Policeman's Beard*.

> *More than iron, more than lead, more*
> *than gold I need electricity.*
> *I need it more than I need lamb or*
> *pork or lettuce or cucumber.*
> *I need it for my dreams.*
>
> —RACTER

---

3. The IRS has indeed developed algorithms to flag "suspicious returns."

Programmer William Chamberlain claims in his introduction that the book contains "prose that is in no way contingent upon human experience." This claim is utterly suspect; every possible aspect of the above "More than iron" poem, for instance, represents the human notion of meaning, of grammar, of aesthetics, even of what a computer might say if it could express itself in prose. Wittgenstein famously said, "If a lion could speak, we could not understand him." Surely the "life" of a computer is far less intelligible to us than that of a lion, on biological grounds; the very intelligibility of Racter's self-description begs scrutiny.

Its structure and aesthetics, too, raise doubts about the absence of a human hand. The anaphora of the first sentence ("more . . . more . . . more") is brought into a nice symmetry with the polysyndeton ("or . . . or . . . or") of the second. The lines also embody the classic architecture of jokes and yarns: theme, slight variation, punch line. These are human structures. My money—and that of many others—says Chamberlain hard-coded these structures himself.

A close structural reading of the text raises important questions about Racter's authorship, as does asking whether the notion of English prose severed from human experience is even a comprehensible idea. But setting these aside, the larger point might be that *no* "demonstration" is impressive, in the way that no prepared speech will ever tell you for certain about the intelligence of the person reciting it.

Some of the earliest questions that come to mind about the capacities of chatbots are things like "Do they have a sense of humor?" and "Can they display emotions?" Perhaps the simplest answer to this type of question is "If a novel can do it, they can do it." A bot can tell jokes—because jokes can be written for it and it can display them. And it can convey emotions, because emotion-laden utterances can be written in for it to display as well. Along these lines, it can blow your mind, change your mind, teach you something, surprise you. But it doesn't make the *novel* a person.

In early 2010, a YouTube video appeared on the Internet of a man having a shockingly cogent conversation with a bot about Shakespeare's *Hamlet*. Some suspected it might herald a new age for chatbots, and for AI. Others, including myself, were unimpressed. Seeing sophisticated behavior doesn't necessarily indicate a *mind*. It might just indicate a *memory*. As Dalí so famously put it, "The first man to compare the cheeks of a young woman to a rose was obviously a poet; the first to repeat it was possibly an idiot."

For instance, three-time Loebner Prize winner Richard Wallace recounts an "AI urban legend" in which "a famous natural language researcher was embarrassed . . . when it became apparent to his audience of Texas bankers that the robot was consistently responding to the *next* question he was about to ask . . . [His] demonstration of natural language understanding . . . was in reality nothing but a simple script."

No demonstration is ever sufficient.

Only *inter*action will do.

We so often think of intelligence, of AI, in terms of *sophistication* of behavior, or *complexity* of behavior. But in so many cases it's impossible to say much with certainty about the program itself, because there are any number of different pieces of software—of wildly varying levels of "intelligence"—that could have produced that behavior.

No, I think sophistication, complexity of behavior, is not it at all. Computation theorist Hava Siegelmann offhandedly described intelligence as "a kind of sensitivity to things," and all of a sudden it clicked—that's it! These Turing test programs that hold forth, these prefabricated poem templates, may produce interesting output, but they're *static*, they don't *react*. They are, in other words, *insensitive*.

Deformation as Mastery

In his famous 1946 essay "Politics and the English Language," George Orwell says that any speaker repeating "familiar phrases" has "gone some distance towards turning himself into a machine." The Turing test would seem to corroborate that.

UCSD's computational linguist Roger Levy: "Programs have gotten relatively good at what is actually said. We can devise complex new expressions, if we intend new meanings, and we can understand those new meanings. This strikes me as a great way to break the Turing test [programs] and a great way to distinguish yourself as a human. I think that in my experience with statistical models of language, it's the unboundedness of human language that's really distinctive."[4]

Dave Ackley offers very similar confederate advice: "I would make up words, because I would expect programs to be operating out of a dictionary."

My mind on deponents and attorneys, I think of drug culture, how dealers and buyers develop their own micro-patois, and how if any of these idiosyncratic reference systems started to become too standardized—if they use the well-known "snow" for cocaine, for instance—their text-message records and email records become much more legally vulnerable (i.e., have less room for deniability) than if the dealers and buyers are, like poets, ceaselessly inventing. A dead metaphor, a cliché, could mean jail.[5]

---

4. That my spell-checker balks at the word "unboundedness" rather poetically demonstrates his point.

5. Yet it's odd—in *other* domains, talking idiosyncratically, freshly, with novel metaphors, makes one *more* easily incriminated. It's easier for someone to find something you said in an email by searching their inbox if you used an *unusual* turn of phrase or metaphor. Things spoken aloud, too, are likely more easily remembered the more unusual and distinctive they are. What's more, in a their-word-against-yours situation, this quotation is likely to be regarded (e.g., by a jury) as more reliable the more unusual and vivid it is.

The general principle, vis-à-vis culpability, would seem to be something along the lines of: if you can obscure your meaning by speaking non-standardly, do so; if your meaning will be clear, speak as generically as possible so as not to be memorable. Writing's goals might be the other way around: clarity with novel ideas and novelty with familiar ones.

In his 1973 book, *The Anxiety of Influence,* Harold Bloom argues that every poet has to, aesthetically, bump off their greatest teacher and influence to become great. To think of language this way brings huge implications for the Turing test. Take even the bots that learn from their human users: Cleverbot, for instance. At *best* it mimics language. It doesn't, as Ezra Pound said, "make it new."

As Garry Kasparov explains in *How Life Imitates Chess,* "In chess, a young player can advance by imitating the top grandmasters, but to challenge them he must produce his own ideas." That is, one can get almost to the top of the chess world—the top two hundred players in the world, say—by merely *absorbing* opening theory. But to crack into the ranks above it requires a player to be challenging that very received wisdom—which all of those players take as a given. To play at that level, one must be *changing* opening theory.

Pound referred to poetry as "original research" in language. When I think about how one might judge the world's best writers, I keep gravitating to the idea that we'd want to look at who changed the language the most. You can barely speak without uttering Shakespeare coinages, like "bated breath," "heart of hearts," "good riddance," "household words," "high time," "Greek to me," "live-long day," the list goes on.

I wonder if bots will be able to pass the Turing test before they make that "transition from imitator to innovator," as Kasparov puts it—before they begin not merely to follow but to lead. Before they make a *contribution* to the language. Which most of us don't think about, but it's part of what we do. "The highest form of warfare is attacking strategy itself," says Sun Tzu. The great chess players change the game; the great artists change their mediums; the most important places, events, and people in our lives change us.

As it turns out, though, you don't have to be Shakespeare to change your language. In fact, quite the opposite: if meaning lies even partially in usage, then you subtly alter the language every time you use it. You couldn't leave it intact if you tried.

Treadmills

"Retarded" used to be a polite word; it was introduced to replace "idiot," "imbecile," and "moron," which had once, themselves, been polite terms. Linguists call this process the "euphemism treadmill." Ironically, to use "retarded" as a way of disparaging a person or idea is more offensive than to use "imbecilic" or "moronic": terms ditched—for being too offensive—in *favor* of "retarded." This lexical switch obviously has not succeeded in the long term. When White House Chief of Staff Rahm Emanuel in a 2009 strategy meeting dubbed a proposal that displeased him "retarded," it prompted calls among prominent Republicans for his resignation, and (in lieu of resignation) a personal apology to the chairman of the Special Olympics. In May of 2010, the Senate's Health, Education, Labor, and Pensions Committee approved a bill called Rosa's Law that would strike "retarded" from all federal language, replacing it with "intellectually disabled." The treadmill continues.

A similar process happens in reverse—the "dysphemism treadmill"—with offensive words; they gradually lose their harshness, and have to be replaced every so often with new abrasives. Some words that are today considered perfectly acceptable, even endearingly inoffensive, or quaint—e.g., "scumbag"—were originally quite explicit: in its original usage, the term meant "condom." As recently as 1998 the *New York Times* was still refusing to print the word, as in, "Mr. Burton's staff today defended his comments, including the use of a vulgarity for a condom to describe the President." But increasing numbers of readers, unaware of the term's etymology—in fact, only a handful of modern *dictionaries* include reference to condoms in the word's definition—were left scratching their heads. By 2006, just eight years later, the paper nonchalantly included the word in a crossword puzzle (clue: "Scoundrel"), prompting outrage—but only among few. The word's origins were news even to puzzle editor and renowned word-guru Will Shortz: "The thought never crossed my mind this word could be controversial."

Other treadmills exist: for instance, slang terms and baby names. Slang invented by an insider group gets picked up by outsider groups, creating the persistent need for new slang to reinforce the insider group's cohesion. In *Freakonomics,* economist Steven Levitt chronicles the process by which baby names percolate through society, from the upper economic classes to the lower ones. Parents often want their child's name to have a ring of success or distinction, and so they look to the names of slightly more successful families; however, this very process begins to deplete the name's cachet, and so the demand shifts gradually and perpetually onto new "high-end" names.

Linguist Guy Deutscher charts two others in *The Unfolding of Language.* The first is the perpetual pull of eloquence and push of efficiency. As he notes, the phrase "up above" has been compacted and elaborated so many times that its etymology is the hopelessly redundant "up on by on up," and likewise some French speakers now say "au jour d'aujourd'hui": "on the day of on-the-day-of-this-day." The second is the constant invention of new metaphors to capture new facets of human experience in language—meanwhile, familiar metaphors are passing, through sheer use, from aptness to popularity to cliché. From there, the fact that it *is* a metaphor is slowly forgotten, the original image at the heart of the term becoming a mere fossil of etymology. For instance, Latin speakers needed a term to describe the relationship they had with their dining partners, the folks they broke bread with: the custom of simply calling such people one's "with-breads," or (in Latin) "com-panis," caught on, a phrasing that eventually became our word "companion." Likewise, as misfortunate events were believed in the sixteenth century to have astrological roots, speakers of Old Italian took to calling such an event a "bad-star," or "dis-astro": hence "disaster."

The language is constantly dying, and constantly being born. English poet John Keats asked that his tombstone read simply, "Here lies one whose name was writ in water": a comment on the ephemerality of life. In the long run, *all* writing is in water: the language itself changes steadily over time. All texts have a half-life of intelligibility before they must be resuscitated through translation.

Language will never settle down, it will never stabilize, it will never find equilibrium. Perhaps part of what makes the Turing test so tricky is that it's a battle on shifting ground. Unlike chess, with its fixed rules and outcomes, language—ever changing—is not amenable to being "solved." As ELIZA's creator, Joseph Weizenbaum, writes, "Another widespread, and to me surprising, reaction to the ELIZA program was the spread of a belief that it demonstrated a general solution to the problem of computer understanding of natural language. [I have] tried to say that no general solution to that problem [is] possible, i.e., . . . even people are not embodiments of any such general solution."

The Observer Effect

You can't take the temperature of a system without the thermometer itself becoming part of the system and contributing its *own* temperature, in some degree, to its reading. You can't check a tire's pressure without letting some of that pressure out—namely, into the gauge. And you can't check a circuit without some of its current flowing into the meter, or vice versa. As Heisenberg famously showed, measuring an electron's position, by bouncing a photon off of it, perturbs the very thing you sought to measure through the act of measurement. Scientists call this the "observer effect."

Likewise, you can't ask a friend if they'd like to go out to dinner without implying the extent to which *you* want to go out to dinner, and thus biasing their answer. Polling studies and eyewitness testimony studies show that the wording of questions biases someone's response—"About how fast were the cars going when they collided into each other?" produces lower estimates than "About how fast were the cars going when they smashed into each other?" Asking

"Do you approve of the job the president is doing?" receives many more affirmatives than asking "Do you approve of the job the president is doing, or not?" The *order* of questions matters too: asking someone about their overall life satisfaction and then their financial satisfaction produces some limited degree of correlation, but asking *first* about their finances and *then* about their life overall magnifies that correlation tremendously.

Computer programming is largely based on the "repeatability" of its responses; as most programmers can attest, a bug that is unrepeatable is also for the most part unfixable. This is part of why computers behave so much better after a reboot than after days of continuous use, and so much better when first purchased than after several years. These "blank-slate" states are the ones most encountered, and therefore most polished, by the programmers. The longer a computer system is active, the more unique its state tends to become. By and large this is true of people too—except people can't reboot.

When I debug a program, I expect to re-create the exact same behavior a number of times, testing out revisions of the code and undoing them where necessary. When I query a computer system, I expect not to alter it. In contrast, human communication is irrevocable. Nothing can be unsaid. (Consider a judge laughably asking the jury to "forget" a piece of testimony.) It is also, in this way, unrepeatable—because the initial conditions can never be re-created.

"Hold still, lion!" writes poet Robert Creeley. "I am trying / to paint you / while there's time to." Part of what I love so much about people is that they never hold still. As you are getting to know them, they are changing—in part due to your presence. (In conversation with one of these PARRY-style personalities, or reading Racter, or watching a video demonstration of a bot in action, I have quite the inverse feeling. I can't get the damn thing to *move*.) In some sense my mind goes to Duchamp's *Nude Descending a Staircase, No. 2*—a series of quick, overlapping sketches of a thing in motion, creating a kind of palimpsest, which scandalized a public accustomed to realism in portraiture. "An explosion in a shingle factory," wrote the appalled

New York Times critic Julian Street, and the piece became a lightning rod for outrage and mockery alike. Somehow, though, there seems something profoundly true (and "realistic") about a human subject that refuses to sit still for the painter, who must aim to capture their essence via *gait* rather than *figure*.

Part of what we have invented the Turing machine–style digital computer for is its reliability, its repeatability, its "stillness." When, in recent years, we have experimented with "neural network" models, which mimic the brain's architecture of massive connectivity and parallelism instead of strict, serial, digital rule following, we have still tended to keep the neuron's amazing plasticity in check. "When the [synaptic] weights [of a network of virtual neurons] are considered constant (after or without a process of adaptation) the networks can perform exact computations," writes Hava Siegelmann. Virtual neurons can be controlled in this way, with strict periods of time in which they are permitted to change and adapt. The human brain has no such limits, owing to what neuroscientists call "synaptic plasticity." Every time neurons fire, they alter the structure of their connections to one another.

In other words, a functioning brain is a changing brain. As Dave Ackley puts it, "You *must* be impacted by experience, or there is no experience." This is what makes good conversations, and good living, *risky*. You can't simply "get to know" a person without, to some degree, changing them—and without, to some degree, becoming them.

I remember first coming to understand that owing to the electrically repellant properties of the atom, matter can never actually *touch* other matter. The notion brought with it the icy chill of solipsism: the self as a kind of hermetically sealed tomb.

Sort of. Someone may not ever quite be able to get to the outside of you. But it doesn't take much—merely to be perceived, or thought of, alters the other's brain—to make it *inside,* to where the self is, and change something there, however small.

Another way to think about it, as you levitate and hover around the room on an angstrom-thick cushion of electromagnetic force, is this: You will never touch anything, in the sense that the nuclei of your arm's atoms will never knock against the nuclei of the table's—for whatever that would be worth. What *feels* like "contact" is actually your body's atoms exerting electromagnetic forces on the table's atoms, and vice versa. In other words, what appears to be static contact is actually dynamic interaction, the exchange of forces.

The very same forces, by the way, that your body's atoms are exchanging with each other—the ones that make you whole.

The Origin of Love

Methinks I have a plan which will humble their
pride and improve their manners; men shall con-
tinue to exist, but I will cut them in two . . .

—ZEUS, QUOTED IN PLATO'S *SYMPOSIUM*

You know we're two hearts
living in just one mind . . .

—PHIL COLLINS

Most folks familiar with Plato's *Symposium*—or, alternately, John Cameron Mitchell's *Hedwig and the Angry Inch*—know Aristophanes' story for the origin of love. People began as eight-limbed creatures: four arms, four legs, two faces. In order to cut us down to size—literally—for our haughtiness at the gods or some such offense, Zeus splits us in two with lightning, and cinches the separated skin together at the belly button, and voilà: humans as we know them today, with just a pair of arms and pair of legs apiece. Out of an ancient need to return to that pre-lightning state of wholeness, we fall in love.[6] All

---

6. Lest you think that this original separation is what created the two sexes, male and female, and that only straight folks have the right ideas about reassembly, remember that Aristophanes, like many Greek men of his time, was

trying to get back to that original whole. The tangle of hugging, kissing, and sex being the closest we can come to "reuniting our original nature, making one of two, and healing the state of man."[7]

As a middle schooler barreling into early adolescence, I used to sit transfixed by late-night MTV screenings of the Spice Girls singing in various states of undress about when "2 Become 1." When we talk about this notion in the context of love, we most frequently mean it as a euphemism for sex. I sometimes think about sex in such Aristophanic terms: a kind of triumphant, tragic attempt to combine two bodies, smooshing them together like clay. Triumphant because it's as close as you ever get.

Tragic for the same reason. Sex never quite seems, in the Aristophanic sense, to *work*—the two never quite manage to become one, and in fact sometimes end up creating a third in the process. Maybe the corporeal reunion, the undoing of Zeus's separation, is simply impossible.[8] When two people marry, there's a *legal* sense in which they "become one"—if only for tax purposes. That, too, though, is hardly the kind of state-of-man repair that Aristophanes imagined.

But there's hope.

---

more *homo-* than hetero-normative. As he explains it, the "sexes were not two as they are now, but originally three in number," corresponding to male, female, and "androgynous"; the male beings when split became gay men, the female beings became lesbians, and the androgynous beings became straight men and women. (No word on how bisexuals fit into this picture.)

7. Most literary metaphors for romantic and sexual passion lean in one way or another toward the violent. We talk about a "stormy" romance or "tempestuous" feelings, or the orgasm as a tiny death (*la petite mort,* the French call it), or of "ravishing" beauty—checking my dictionary, I see that "ravishing" as an adjective means charming or gorgeous, and as a noun or verb, rape. And most slang terms for sex are violent—bang, screw—or at the very least negative. It's hard to imagine ending up in better shape than when you started. But for Aristophanes it wasn't violence at all, but *healing*—it's no wonder his is such an endearing (and enduring) myth.

8. Yet I think of Sean Penn's answer, in *Milk,* to the question of whether men can reproduce: "No, but God knows we keep trying."

Nervous System to Nervous System: Healed by Bandwidth

The organizer of the 2008 Loebner Prize was University of Reading professor Kevin Warwick—also known in the press sometimes as "the world's first cyborg." In 1998 he had an RFID chip implanted in his arm: when he walks into his department, the doors open for him and a voice says, "Hello, Professor Warwick." More recently, he's undergone a second surgery, a much more invasive one: wiring a hundred-electrode array directly into the nerves of his arm.

With this array he's done a number of equally astonishing things: he's been able to get a robot arm to mimic the actions of his real arm, using the electrode array to broadcast the neural signals coming from his brain to the robot arm, which follows those commands in real time, just as—of course—Warwick's real arm does.[9]

He also experimented with adding a sixth sense—namely, sonar. A sonar device attached to a baseball cap routed its signals into Warwick's arm. At first, he says, he kept feeling as though his index finger was tingling whenever large objects got near him. But in very little time, the brain accustomed itself to the new data and the sensation of finger tingling went away. Close-by objects simply produced a kind of ineffable "oh, there's an object close by" feeling. His brain had made sense of and integrated the data. He'd acquired a sixth sense.

One of the landmark philosophy of mind papers of the twentieth century is Thomas Nagel's 1974 "What Is It Like to Be a Bat?" Well, as far as sonar's concerned, there's one man alive who might actually be able to hazard an answer to Nagel's famously unanswerable, and largely rhetorical, question.

Perhaps the most amazing thing that Warwick did with his arm

---

9. I suppose I shouldn't say "of course": there was actually a serious risk that the surgery would leave Warwick paralyzed. Somehow that didn't seem to faze him.

socket, though, is what he tried next. Warwick wasn't the only one to get silicon grafted into his arm nerves. So did his wife.

She would make a certain gesture with her arm, Warwick's arm would twinge. Primitive? Maybe, yes. But Warwick waxes Wright brothers/Kitty Hawk about it. The Wrights were only off the ground for seconds at first; now we are used to traveling so far so fast that our bodies get out of sync with the sun.[10]

A twinge is ineloquent: granted. But it represents the first direct nervous-system-to-nervous-system human communication. A signal that shortcuts language, shortcuts gesture.

"It was *the* most exciting thing," says Warwick, "I mean, when that signal arrived and I could understand the thing—and *realizing* what potentially that would mean in the future— Oh, it was the most exciting thing by far that I've been involved with."[11]

What *might* it mean in the future? What might the Lindbergh- or Earhart-comparable voyage be? As Douglas Hofstadter writes, "If the bandwidth were turned up more and more and more and still more . . . the sense of a clear boundary between them would slowly be dissolved."

Healed at last? By *bandwidth,* of all things? It's not as crazy as it sounds. It's what's happening right now, in your own head.

The Four-Hemisphere Brain

Our uniquely human skills may well be produced by
minute and circumscribed neuronal networks. And
yet our highly modularized brain generates the feeling

---

10. The Conestoga wagoners, for instance, taking six months to make the trip I cram into the evening before Thanksgiving, didn't seem to have this problem.
11. "You're getting on my nerves," we imagine him saying, suggestively. "Oh, you're such a tase," she, atingle, replies . . .

in all of us that we are integrated and unified. How so,
given that we are a collection of specialized modules?

—MICHAEL GAZZANIGA

The only sex relations that have real value are
those in which there is no reticence and in
which the whole personality of both becomes
merged in a new collective personality.

—BERTRAND RUSSELL

What Warwick and Hofstadter are talking about is not nearly so fantastical or sci-fi as it sounds. It's a part of the brain's very architecture, where the several hundred million fibers of the corpus callosum are ferrying information back and forth between our twin organs of thought, our left and right hemispheres, at an extremely high—but finite—rate. Set lovers aside for a moment: the integrity and coherence of the *mind,* the oneness of the *self,* is dependent on data transfer. On communication.

One metaphysical oddity: communication comes in *degrees.* The number of *minds,* the number of *selves,* in a body, seemingly, doesn't. This begs odd questions. If the bandwidth of one's corpus callosum were turned up just slightly, would that make someone somehow "closer" to *one* self? If the bandwidth were turned down just slightly, would that make someone somehow *farther* from one self? With the bandwidth right where it is now, how many selves *are* we, exactly?[12]

This intense desire to make one of two, to be "healed" and restored

---

12. As it turns out, "axonal diameter" (thicker neurons signal faster over long distances but take up more space) correlates with brain size for virtually all animals, *except*—as neurophysiologist Roberto Caminiti recently discovered—humans. Our axonal diameter is not significantly greater than chimpanzees', despite our having larger brains. Evolution appears to have been willing to trade interhemispheric lags for a disproportionate increase in computational power.

to unity: this is the human condition. Not just the state of our sexuality, but the state of our minds. The eternal desire to "catch up," to "stay connected," in the face of flurrying activity and change. You never really gain ground and you never really lose ground. You aren't unified but you aren't separate.

"They're basically the same person," we sometimes say of a couple. We may not be entirely kidding. There's a Bach wedding cantata where the married couple is addressed with second-person-*singular* pronouns. Because in English these are the same as the second-person-plural pronouns—"you," "your"—the effect doesn't quite translate. However, we do sometimes see the opposite, where a coupled partner describes events that happened only to him- or herself, or only to the partner, using "we"—or, more commonly, simply talks about the couple as a *unit,* not as "she and I." A recent study at UC Berkeley, led by psychology Ph.D. student Benjamin Seider, found that the tendency toward what he calls linguistic "we-ness" was greater in older couples than younger ones.

Considering that the brain itself stays connected only by constant conversation, it's hard to argue that our connections to others belong strictly on a lower tier. What makes the transmissions passing through the corpus callosum all that different from the transmissions passing through the air, from mouth to mouth? The intra-brain connections are *stronger* than the inter-brain connections, but not totally different in kind.

If it's communication that makes a whole of our two-hemisphere brain, there should be no reason why two people, communicating well enough, couldn't create the *four*-hemisphere brain. Perhaps two become one through the same process *one* becomes one. It may end up being talk—the *other* intercourse—that heals the state of man. If we do it right.

10. High Surprisal

One-Sided Conversations

Eager to book my room in Brighton, I did some quick digging around online and found an intriguing (and intriguingly named) place, just a stone's throw from the Turing test, called "Motel Schmotel." I called them up via Skype. Now, I don't know if it was the spotty connection, or the woman's low speaking volume, or the English accent, or what, but I could barely understand a word of what she was saying, and immediately found myself hanging on to the flow of the conversation for dear life:

——*tel*.

Presumably, she's just said something like "Hello, Motel Schmotel." No reason not to plunge ahead with my request.

Uh, yeah, I'd like to check the availability for a single room?

——*ong?*

Probably "For how long?" but hard to know for sure. At any rate, the most likely thing she needs to know if I'm looking for a room is the duration, although that's not helpful without the start date, so why don't I nip that follow-up question (which I probably won't hear anyway) in the bud and volunteer both:

Um, for four nights, starting on Saturday the fifth of September?

——*[something with downward tone]*——*, sorry. We only*—— *balcony*——*ninety pounds.*

And here I was lost. They didn't have *something*, but apparently they had *something* else. Not clear how to proceed. (Ninety pounds total, or extra? Per night, or for the whole stay? I couldn't do all the math in my head at once and figure out if I could afford the room.) So I hedged my bets and said the most utterly neutral, noncommittal thing I could think of:

Ah. Okay.

——rry!

Presumably "sorry," and said with a friendly finality that seemed to signal that she was expecting me to hang up any second: probably this balcony room was out of my league. Fair enough.

Ok, thanks! Bye.

——ye, now!

I suppose I got off the phone with a mixture of bemusement and guilt—it hadn't actually been all that necessary to hear what she was saying. I *knew* what she was saying. The boilerplate, the template of the conversation—my ability to guess at what she was asking me and what her possible responses could be—pulled me through.

On (Not) Speaking the Language

It occurred to me that I'd been able to pull off the same trick the last time I'd been in Europe, a two-week-long Eurail-pass whirlwind through Spain, France, Switzerland, and Italy the summer after college: though I speak only English and some Spanish, I did much of the ticket buying in France and Italy, and managed for the most part to pull it off. Granted, I did nod understandingly (and, I confess, impatiently) when the woman sold us our overnight tickets to Salzburg and kept stressing, it seemed to me unnecessarily, "est . . . station . . . est . . . est . . . station"—"This station, this one, I understand, yeah, yeah," I replied, knowing, of course, that in Spanish "este" means *this*. Of Paris's seven train stations, our overnighter would be coming to *this* one, right here.

Perhaps you're saying, "Wait, Salzburg? But he didn't say anything about seeing Austria . . ." Indeed.

What I failed to remember, of course, is that "este" in Spanish *also* means *east*—a fact that dawns on me as we stand dumbfounded on a ghostly empty train platform at midnight in Paris's Austerlitz Station, checking our watches, realizing that not only have we blown the chance at realizing our Austrian *Sound of Music* Alp-running fantasies, but we'll have to do an emergency rerouting of our entire, now Austria-less, itinerary, as we'll be just about a million meters off course by morning. Also: it's now half past midnight, the guidebook is describing our present location as "dicey," and we don't have a place to sleep. Our beds have just departed from *East Station,* fast on their way to Salzburg.

Now, this rather—ahem—serious exception aside, I want to emphasize that by and large we did just fine, armed in part with my knowledge of a sibling Romance language, and in part with a guidebook that included useful phrases in every European tongue. You realize the talismanic power of language in these situations: you read some phonetically spelled gobbledygook off a sheet, and before you know it, abracadabra, beers have appeared at your table, or a hostel room has been reserved in your name, or you've been directed down a mazelike alley to the epicenter of late-night flamenco. "Say the magic word," the expression goes, but all words seem, in one way or another, to conjure.

The dark side of this is that the sub-fluent traveler risks solipsism—which can only be cracked by what linguists and information theorists call "surprisal," which is more or less the fancy academic way of saying "surprise." The amazing thing about surprisal, though, is that it can actually be *quantified* numerically. A very strange idea—and a very important one. We'll see how exactly that quantification happens later in this chapter; for now, suffice it to say that, intuitively, a country can only become real to you, that is, step out of the shadow of your stereotypes of the place, by surprising you. Part of this requires that you *pay attention*—most of life's surprises are on the small side and often go unnoticed. The other part of it requires that you put yourself into situations where surprise is possible; sometimes this requires merely the right attitude of openness on your part, but other times it's impossible without serious effort and commitment ahead of time (e.g., learning the language). Template-based interactions—"Je voudrais un hot dog, s'il vous plaît . . . merci!"; "Où est le WC? . . . merci!"—where you more or less treat your interlocutor as a machine, are navigable for precisely the reason that they are of almost no cultural or experiential value. Even if your interlocutor's response is surprising or interesting, you might miss it. *Wielding* language's magic is intoxicating; becoming *susceptible* to it, even more so.

Perhaps you're starting to feel by now how all of this parallels the Turing test. In France I behaved, to my touristy chagrin, *like a bot*. *Speaking* was the easy part—provided I kept to the phrase book (this in itself was embarrassing, that my desires were so similar to those of every other American tourist in France that a one-size-fits-all FAQ sheet sufficed handily). But *listening* was almost impossible. So I tried only to have interactions that didn't really require it.

Interacting with humans in this way is, I believe, shameful. The Turing test, bless it, has now given us a yardstick for this shame.

A Mathematical Theory of Communication

It seems, at first glance, that information theory—the science of data transmission, data encryption, and data compression—would be mostly a question of engineering, having little to do with the psychological and philosophical questions that surround the Turing test and AI. But these two ships turn out to be sailing quite the same seas. The landmark paper that launched information theory is Claude Shannon's 1948 "A Mathematical Theory of Communication," and as it happens, this notion of scientifically evaluating "communication" binds information theory and the Turing test to each other from the get-go.

What is it, exactly, that Shannon identified as the essence of communication? How do you *measure* it? How does it help us, and how does it hurt us—and what does it have to do with being human?

These connections present themselves in all sorts of unlikely places, and one among them is your phone. Cell phones rely heavily on "prediction" algorithms to facilitate text-message typing: guessing what word you're attempting to write, auto-correcting typos (sometimes overzealously), and the like—this is data compression in action. One of the startling results that Shannon found in "A Mathematical Theory of Communication" is that text prediction and text *generation* turn out to be mathematically equivalent. A phone that could consistently anticipate what you were intending to write, or at least that could do as well as a human, would be just as intelligent as the program that could *write you back* like a human. Meaning that the average American teenager, going by the *New York Times*'s 2009 statistics on cell phone texting, participates in roughly eighty Turing tests a day.

This turns out to be incredibly *useful* and also incredibly *dangerous*. In charting the links between data compression and the Turing test's hunt for the human spark, I'll explore why. I want to begin with a little experiment I did recently, to see if it was possible to use a computer to quantify the literary value of James Joyce.[1]

James Joyce vs. Mac OS X

I took a passage from *Ulysses* at random and saved it on my computer as raw text: 1,717 bytes.

Then I wrote the words "blah blah blah" over and over until it matched the length of the Joyce excerpt, and saved that: 1,717 bytes.

Then I had my computer's operating system, which happens to be Mac OS X, try to compress them. The "blah" file compressed all the

---

1. Claude Shannon: "Joyce . . . is alleged to achieve a compression of semantic content."

way down to 478 bytes, just 28 percent of its previous size, but *Ulysses* only came down to 79 percent of its prior size, or 1,352 bytes—leaving it nearly three times as large as the "blah" file.

When the compressor pushed down, something in the Joyce pushed back.

Quantifying Information

Imagine flipping a coin a hundred times. If it's a fair coin, you can expect about fifty heads and fifty tails, of course, distributed randomly throughout the hundred. Now, imagine telling someone which flips came out which—it'd be a mouthful, of course. You could name all of the outcomes in a row ("heads, heads, tails, heads, tails, . . .") or just the location of either just the heads ("the first, the second, the fourth, . . .") or just the tails, letting the other be implicit, both of which come out to be about the same length.[2]

But if it's a biased coin, your job gets easier. If the coin comes up heads only 30 percent of the time, then you can save breath by just naming which flips came up heads. If it's heads 80 percent of the time, you simply name which flips were *tails*. The more biased the coin, the easier the description becomes, all the way up to a completely biased coin, our "boundary case," which compresses down to a single word—"heads" or "tails"—that describes the entire set of results.

So, if the result of the flips can be expressed with less language the more biased the coin is, then we might argue that in these cases the result literally contains less *information*. This logic extends down, perhaps counterintuitively, perhaps eerily, into the individual events themselves—for any *given* flip, the more biased the coin, the less information the flip contains. There's a sense in which flipping the

---

2. Length here refers to binary bits, not words of English, but the distinction isn't hugely important in this case.

seventy-thirty coin just doesn't deliver what flipping the fifty-fifty coin does.[3] This is the intuition of "information entropy": the notion that the amount of information in something can be measured.

"Information can be measured"—at first this sounds trivial, of course. We buy hard drives and fill them up, wonder if shelling out the extra fifty dollars for the 16 GB iPod will be worth it compared to the 8 GB one, and so on. We're used to files having size values in bytes. But the size of a file is not the same thing as the amount of information in a file. Consider as an analogue the difference between volume and mass; consider Archimedes and the golden crown—in order to determine whether the crown's gold was pure, Archimedes needed to figure out how to compare its mass to its volume.[4] How do we arrive at the density of a *file*, the karat of its bytes?

Information, Bias, and the Unexpected

We could compress the biased coin *because* it was biased. Crucially, if all outcomes of a situation are equally probable—what's called a "uniform distribution"—then entropy is at its maximum. From there it decreases, all the way to a minimum value when the outcome is fixed or certain. Thus we might say that as a file hits its compres-

---

3. This is why, for instance, starting a game of Guess Who?, as I routinely did in the late '80s, by asking about the person's gender is a poor strategy: the game only had five women characters to nineteen men, so the question wasn't as incisive as one that would create a twelve-twelve split.

4. The problem was how to get an accurate gauge of its volume without melting it down. Thinking about this as he stepped into a public bath, all of a sudden he registered: the water level rose as he got in! You can measure the volume of an irregular object by the amount of water it displaces! Allegedly, he was so excited about this insight that he immediately leaped out of the bath and ran home to work out the experiment, naked and dripping bathwater through the streets, shouting for joy. The word he was shouting was the Greek for "I've got it!" and has since become our synonym for scientific discovery: *Eureka.*

sion floor, the fixities and certainties shake out; pattern and repetition shake out; predictability and expectancy shake out; the resulting file—before it's decompressed back into its useful form—starts looking more and more random, more and more like white noise.

Information, defined intuitively and informally, might be something like "uncertainty's antidote." This turns out also to be the formal definition—the amount of information comes from the amount by which something reduces uncertainty. (Ergo, compressed files look random: nothing about bits 0 through n gives you any sense what bit $n + 1$ will be—that is, there is no pattern or trend or bias noticeable in the digits—otherwise there would be room for further compression.[5]) This value, the informational equivalent of mass, comes originally from Shannon's 1948 paper and goes by the name of "information entropy" or "Shannon entropy" or just "entropy."[6] The higher the entropy, the more information there is. It turns out to be a value capable of measuring a startling array of things—from the flip of a coin to a telephone call, to a Joyce novel, to a first date, to last words, to a Turing test.

The Shannon Game

One of the most useful tools for quantitatively analyzing English goes by the name of the Shannon Game. It's kind of like playing hangman, one letter at a time: the basic idea is that you try to guess the letters of a text, one by one, and the (logarithm of the) total number of guesses

---

5. As a result, highly compressed files are much more fragile, in the sense that if any of the bits are corrupted, the context won't help fill them in, because those contextual clues have already been capitalized on and compressed away. This is one of the useful qualities of redundancy.

6. Not to be confused with *thermodynamic* entropy, the measure of "disorder" in a physical system. The two are in fact related, but in complicated and mathematically strenuous ways that are outside of our scope here but well worth reading about for those curious.

required tells you the entropy of that passage. The idea is to estimate how much knowledge native speakers bring to a text. Here's the result of a round of the Shannon Game, played by yours truly:[7]

```
U N D E R N E A T H _ T H E _ B L U E _
22 1 1 1 1 1 1 1 1 1 1 2 1 1 1 5 6 5 1 2

C U S H I O N _ I N _ T H E _ L I V I N G _
2 7 11 5 1 1 1 2 6 5 2 1 1 1 1 1 1 1 1 1 1

R O O M _ I S _ A _ H A N D F U L _ O F _
1 1 1 1 1 1 1 1 1 19 3 1 2 13 5 1 1 1 1 1

C H A N G E _ A N D _ T H E _ R E M O T E _
1 21 1 1 2 1 1 1 1 1 1 6 1 1 1 4 2 9 5 1 1 1

C O N T R O L
1 1 1 1 1 1
```

We can see immediately that the information entropy here is wildly nonuniform: I was able to predict "the_living_room_is_a_" completely correctly, but almost exhausted the entire alphabet before getting the *h* of "handful"—and note how the "and" in "handful" comes easily but the entropy spikes up again at the *f*, then goes back down to the minimum at *l*. And "remo" was all I needed to fill in "te_control."

---

7. Play the game yourself at math.ucsd.edu/~crypto/java/ENTROPY/. It's fun; plus, moving that slowly and being forced to speculate at every single darn step of the way, you'll never think about language and time the same way again. Some elementary schools use a variation of the Shannon Game to teach spelling; I'd have my undergraduate poetry workshop students play the Shannon Game to strengthen their syntactical chops. In poems, where the economy of language is often pushed to its breaking point, having a feel for what chains of words will be predictable to a reader is a useful compass.

Search and the Shannon Game

We computer users of the twenty-first century are perhaps more aware of information entropy—if not by name—than any generation before. When I use Google, I intuitively type in the most unusual or infrequent words or phrases, ignoring more common or expected words, as they won't much narrow down my results. When I want to locate a passage in the huge MS Word document that contains this manuscript, I intuitively start to type the most unusual part of the passage I have in mind: either a proper noun or an unusual diction choice or a unique turn of phrase.[8] Part of the effectiveness of the strange editing mark "tk," for "to come," is that *k* rarely follows *t* in English, much more rarely than *c* follows *t,* and so a writer can easily use a computer to sift through a document and tell him whether there are any "tk's" he missed. (Searching this manuscript for "tc" pulls up over 150 red herrings, like "watch," "match," throughout the manuscript; but with only one exception, all occurrences of "tk"—out of the roughly half-million characters that comprise a book—appear in this paragraph.) When I want to pull up certain songs or a certain band in my iTunes library, say Outkast, I know that "out" is such a prevalent string of letters (which pulls up all Outkast songs, plus 438 others I don't want) that I'm better off just typing "kast" into the search box. Or even just that same trusty rare bigram "tk," which pulls up all the songs I want and only *three* I don't.

Art and the Shannon Game

"Not-knowing is crucial to art," says Donald Barthelme, "is what permits art to be made." He's referring here to the *what happens if I try this?* and *what do I do next?* of the creative process, but I think it's

---

8. Fascinatingly, this suggests that blander, more generic, lower-vocabulary, or more repetitive books are harder to search, and harder to edit.

just as true a statement about what it's like to be a *reader.* "Every book, for me, is the balance of YES and NO," writes one of Jonathan Safran Foer's narrators in *Extremely Loud and Incredibly Close.* The Shannon Game represents one approach, one very fine-grained approach, to thinking about the reading experience as a kind of extremely rapid sequence of guesses, and much of the satisfaction, it would seem, is in the balance between yes and no, affirmation and surprise. Entropy gives us a quantifiable measure of where *exactly* the not-knowing lies, how *exactly* the YES and NO congregate on the page. Going back to the original spirit of Barthelme's statement, does entropy give us a road into the creative imagination as well? Are the unguessable moments also the most creative ones? My intuition says yes, there's a link. Anecdotally, here's the next round of the Shannon Game I tried:

EVEN_THOUGH_YOU_DONT_KNOW_HOW_TO_

FLY_YOU_MIGHT_BE_ABLE_TO_LIFT_YOUR_

SHOE_LONG_ENOUGH_FOR_THE_CAT_TO_

MOVE_OUT_FROM_UNDER_YOUR_FOOT

The highest-entropy letters (when I attempted this passage) were the *Y* in the first "you," the *C* in "cat," and the *M* in "move." They're also, interestingly, key grammatical moments: respectively, the subject of the first dependent clause and the subject and verb of the second dependent clause. And are these not also the moments where the author's intention and creativity peak too? And are these not the words—especially "cat"—that, if removed, would be the most difficult for the reader to guess back in?

This last measure actually has its own name; it's called the "cloze test." The name comes from something in Gestalt psychology called the "law of closure"—which refers to the observation that when people look at a shape with some missing pieces or erasures, they still in some sense "experience" the parts that are missing.[9] Cloze tests make up some of the familiar SAT-type questions where you have to fill the most context-sensible word into a blank in a _____ (a) pretzel, (b) Scrabble tile, (c) machine-gun, (d) sentence. Remove the context clues and ask the question anyway, and you have one of my favorite sources of amusement as a child: Mad Libs.

Crowded Room as Cloze Test

However, even if one has to sit down to the SAT or open a Mad Libs book to run into the cloze test on *paper,* the *oral* version of the cloze test is so common it's essentially unavoidable. The world is noisy—we are always trying to talk over the sound of the wind, or construction across the street, or static on the line, or the conversations of others (who are themselves trying to talk over *us*). The audible world is a cloze test.

Though it seems academic, the effects of this are visible everywhere. My conversation with the Brighton hotel clerk was a cloze test with the blanks almost the size of the sentences themselves; still I could guess the right answers back in. But—and for this very reason—I wouldn't call it a particularly human interaction. When I think about, say, the most thrilling intellectual discussions I've had with friends, or the best first dates I've had, I can't begin to imagine them holding up with that many blanks to fill in. I wouldn't have kept up.

Also, think about how you talk to someone when there's loud music playing—so often we start to iron out the idiosyncrasies in our diction and phrasing. Whether we've heard of the cloze test, the Shannon Game, and information entropy or not makes no difference: we know intuitively when—and how—to navigate them, and when and how to enable them to be navigated easily by others. "Let's blow this popsicle

---

9. For this reason much swear-bleeping censorship on television makes no sense to me, because if the removed words are cloze-test obvious, then to what extent have you removed them?

stand," I might say if I can be heard clearly. But noise does not toler-
ate such embellishment. ("Let's go get pot stickers, man?" I imagine
my interlocutor shouting back to me, confused.) No, in a loud room
it's just, "Let's go."

It strikes me as incredibly odd, then, that so many of the sites
of courtship and acquaintance making in our society are so *loud*.[10]
There are any number of otherwise cool bars and clubs in Seattle that
my friends and I avoid: you walk in and see small clusters of people,
huddled over drinks, shouting at each other, going hoarse over the
music. I look in and think, as a confederate, the Turing test would be
harder to defend in here. The noise has a kind of humanity-dampening
effect. I don't like it.

Lossiness

There are two types of compression: "lossless" and "lossy." Lossless compression means that nothing is compromised; that is, upon
decompression we can reconstruct the original *in its entirety* without
any risk of getting it wrong or missing anything or losing any detail.
(ZIP archives are one such example, your photos and documents
unharmed by the process of making the archive.) The other type
of compression is what's called *lossy;* that is, we may *lose* some data
or some level of detail as a cost of the compression. Most images
that you see on the web, for instance, are lossy compressions of larger
digital photos, and the MP3 files on your computer and iPod are
lossy compressions of much higher-resolution recordings at the labels.
The cost is a certain amount of "fidelity." tkng ll f th vwls t f ths
sntnc, fr xmpl, nd mkng t ll lwrcs, wld cnsttt lssy cmprssn: for the
most part the words can be reconstructed, but ambiguities arise here
and there—e.g., "swim," "swam," and "swum"; "I took a hike today"

---

10. Heck, even pickup artists don't like it. As Mystery puts it, "The location
where you first encounter a woman is not necessarily favorable . . . The music
may be too loud for lengthy comfort-building dialogue."

and "Take a hike, toady"; "make a pizza" and "Mike Piazza." In many instances, though—as is frequently the case with images, audio, and video—getting an *exact* replication of the original isn't so important: we can get close enough and, by allowing that wiggle room, save a great deal of time, space, money, and/or energy.

Entropy in Your Very Own Body

Lest you think that the Shannon Game's scoring of texts is an abstraction pertinent only to computer scientists and computational linguists, you might be interested to know that Shannon entropy correlates not only to the metrical stresses in a sentence (poets, take note) but also to the pattern by which speakers enunciate certain words and swallow others.[11] So even if you've never heard of it before, something in your head intuits Shannon entropy every time you open your mouth. Namely, it tells you how far to open it.[12]

It turns out, too, that if we map the movements of readers' eyes, their "saccades" and "fixations," the way they dance over the text, their time spent snagging on (or returning to) certain parts of a passage corresponds quite nicely to its Shannon Game values. "Transitional probabilities between words have a measurable influence on fixation durations," write the University of Edinburgh's Scott McDonald and Richard Shillcock. "Readers skipped over predictable words more than unpredictable words and spent less time on predictable words when they did fixate on them," writes a team of psychologists from the University of Massachusetts and Mount Holyoke College.

---

11. That the distinctions between the words in "And in an . . ." are steamrollered by native speakers, especially when talking excitedly—"Nininin . . ."—is lossy compression. We can afford to render all three words alike because the rules of grammar and syntax prevent other "decompressions," like "and and an," or "an in in," from seeming plausible.

12. I can't help noticing that the fifteen-character "Shannon entropy" and five-character "mouth" have both been assigned a two-character substitute—in fact, the *same* two-character symbol, the pronoun "it"—in this sentence: more compression for you.

As researchers Laurent Itti and Pierre Baldi put it, "Surprise explains best where humans look . . . [It] represents an easily computable shortcut towards events which deserve attention." In other words, entropy guides the eye. It gives every passage a secret shape.

Artifacts

Lossy compression brings along with it what are known as compression "artifacts"—the places where the lossiness of the compression process leaves its scars on the data. The interesting thing about compression artifacts is that they are *not* random—as a matter of fact, they have a kind of signature. Two of the primary image formats on the web—GIF and JPEG—each leave a characteristic mark: The JPEG's will be regions of what looks like turbulence or heat distortion in areas that would otherwise be either uniform color or sharply divided between colors and textures. The GIF's will be speckles of one color against a background of a similar color (dithering) or smooth gradients of color divided into strips of uniform color (color banding).

A "computer forensics" researcher named Neal Krawetz has used the compression artifacts in al Qaeda videos—employing a technique called "error level analysis"—to demonstrate that elements of the background are frequently green-screened in. He's used the same techniques to document the startling ways that graphic designers in the fashion industry modify the photographs of their models.

One of the strange artifacts that we're quickly growing subconsciously used to is lag. When you play a DVD on a slow computer, notice that the moments when the scene changes quickly or when the camera is moving swiftly through an environment are the moments when the computer is most likely to start lagging. (For this reason, music videos, action movies, and, ironically, commercials, all with higher than normal rates of camera cuts and/or camera motion, suffer the most from compression. Romantic movies and sitcoms, with their slower cuts and frequent shots of people standing still and talking, hold up better, for instance, when you stream them over the Internet.) The implication is that these moments must contain more information per second of footage than other moments, such as a character talking against a static background. When you play a graphics-intensive computer game, look for the moments when the frame rate—how many fresh updates to the screen the computer can provide per second—suddenly drops. Some MP3 files use "variable bit rate encoding"—their sampling frequency changes depending on how "complex" the song is at each moment. Sound files are in general so much smaller than video files that you're not likely to hear lag at the moments when the bit rate spikes up, but the principle is the same.

Compare this to watching a projected film in the theater: the entire image is swapped out every twenty-fourth of a second. In many shots, much of what's on-screen remains the same or changes very little in that twenty-fourth of a second. So there's a needless waste of energy. But the waste has a freeing effect: the projector and the filmstrip don't need to know or care how dramatically the image is changing, whereas your computer, streaming data over the Internet and trying to milk the most from each bit, is sensitive to those things.

And, because so much of what we see, hear, and do is compressed, we become sensitive to these things too. A live recording gets assigned a variable bit rate; a computer simulation of an event produces a variable frame rate. Could it be that the entropy of life *itself* is uneven? What might a study of compression have to teach us about the business of living well?

Lossiness and Stakes

One of the strange things about loss*less* compression is that certain things turn out to have a counterintuitively high information entropy. One example is static. Because static, both audio and visual, is random, by definition there aren't patterns that a compressor could exploit; thus it has essentially the highest information entropy. What seems strange about this is that the *stakes* of that information are low—how can we have a lot of information, yet none of it worthwhile?

The picture changes, though, when we consider lossy compression. Ultimately, the quality of lossy compressions has to be judged by human eyes and ears—it isn't as exact a science as lossless compression. One thing that does become clear is that certain things can be pushed fairly far before they show subjective quality loss. Here's where "stakes" enter. To "accurately" capture television snow, for instance, one simply needs a general sense of its range of colors and general texture. Because it's so nearly random, it is extremely hard to *losslessly* compress, but will *lossily* compress down to almost nothing.

Lossy compression is fuzzier: it's more subjective and inexact as a field. But it's the far more pervasive kind. Every time you answer a phone call from your mother and she asks you how your day was, the answer she receives will be lossily compressed thrice: by the phone company, shaving away certain measures of audio fidelity to fit more calls on its network bandwidth; by you, turning roughly 60 sec/min $\times$ 60 min/hr $\times$ 8 hrs = 30,000 seconds of lived experience into a couple hundred syllables; and, last, by her own memory, which shucks those syllables to get at a paraphrasable "gist," forgetting most of the syllables themselves within seconds.

It follows that not only excerpt and quotation but *description* is a form of lossy compression. In fact, lossy compression is the very essence of what language *is*. It accounts for both its immense shortcomings and its immense value—and is another example of the "not-knowing" that allows for art.

Compression and Literary Technique

Synecdoche, the linguistic device by which we name a part but mean the whole—"a new set of *wheels*," meaning car; "*mouths* to feed," meaning people; "nice *threads*," meaning clothing item—gives us one option of transmission in which we attempt to preserve and transmit the *most salient* part of an experience, with the understanding that the reader will fill in the rest. The storyteller who uses synecdoche is like a botanist who returns from the field with a stem cutting that grows an entire tree—or the blue Linckia sea star, whose severed arm will generate itself a new body. The piece gives you back the whole.

T. S. Eliot says in his famous 1915 poem "The Love Song of J. Alfred Prufrock," "I should have been a pair of ragged claws / Scuttling across the floors of silent seas." Given the claws, we imagine the rest of the crustacean's body quite clearly, whereas if he'd said "I should have been a ragged crab," we would *know* the claws were there, of course, but they'd be less vivid, fuzzed out, lower resolution.

Similar to synecdoche is the use of "enthymemes," a technique in argumentation where you explain a piece of reasoning but leave out a premise (because it's assumed to be understood) or the conclusion (because you want your audience to derive it on their own). An example of the former would be to say "Socrates is a man, so Socrates must eventually die," where the obvious second premise, "All men must eventually die," is left unstated. Leaving out a premise, when you're confident that your interlocutor can fill it back in, speeds things up and avoids stating the obvious.[13] And leaving off the conclusion can produce drama, leading the audience all the way up to a point but letting them come to it themselves: "Well, Socrates is a man, and all men must eventually die, so . . ." There's some evidence that in courtroom closing statements and classroom lectures, making the

---

13. Any utterance or description or conversation, of course, leaves countless things out. Thus the implication of anything *said* is that it is, in fact, non-obvious. Thus the word "obviously" (or the phrase "of course") is always at least slightly disingenuous—because anything said must be at least *somewhat* surprising and/or informative in order to *be* said. (Everything said has a presumption of ignorance behind it. This is why stating the obvious is not only inefficient, but frequently offensive. Yet the opposite, too much left unsaid—as Shannon shows us in the value of redundancy, and as expressions like "when you assume you make an ass out of u and me" indicate—has its own risks.)

audience (jurors or students) assemble the conclusion or "punch line" themselves is more engaging and therefore makes a greater impact. (This assumes, however, that they arrive at the conclusion you intend. Other conclusions—e.g., corpses *are* ticklish!—may be possible; this is the lossiness of enthymemes.)

Similarly, when we use the technique of "aposiopesis," a technique popular among screenwriters and playwrights where a thought or line of dialogue is suddenly broken off—

Criticism as Compression

You can think of *criticism* as compression too: a work of literature must strain to survive and outlast its own marketing and its own reviews, which threaten, in a sense, to deliver a lossy compression of the book itself. Anything said about a piece of art enters into competition with the art.

People complain from time to time about folks who read the Cliffs-Notes to a book, or reviews or essays about a book, but don't read the book. Hey, if the information density of, say, *Anna Karenina* is low enough that a review 1 percent as long conveys 60 percent of the form and content "gist" of the book, then it's Tolstoy's fault. His readers are human beings with only twenty-eight thousand days or so separating birth and death. If they want to read the lossy gloss and move on, who can blame them?

Likewise for conceptual art: who needs to *see* a Duchamp toilet when you can *hear* about one so much faster and extract most of the experience from that? Conceptual art might be, for better or worse, (definable as) the art most susceptible to lossy compression.

Showing vs. Telling

"Show, don't tell" is the maxim of many a creative writing workshop. Why is that? Well, for one, it's information entropy. When we talk about a missing tooth, we can be led by that single image, in the right context, to imagine an entire bygone childhood era, an entire history of spousal abuse, or—as is the case in the chilling C. D. Wright poem "Tours"[14]—both at once. Whereas being *told* that a spouse has long been abused, or that a daughter is growing up, might not get us to imagine something as specific and vivid as the missing tooth.

But, as an argument for showing over telling, this line of thinking shouldn't be allowed to become dogma; it's an *empirical* question, ultimately. There are indeed times when the information entropy of telling exceeds that of showing. When we as writers or as speakers encounter them, we need to bend to the higher rule.

An author who has mastered this is Milan Kundera. When he needs to "say" something to the reader in one of his novels, he doesn't construct an elaborate pantomime in which his characters, interacting with each other, subtly convey it: rather, he, Kundera, just steps in and says it. ("As I pointed out in Part One . . .") How sublime! Imagine a street mime giving up on the exasperating charades and saying, simply, "I'm trapped in a box."

Entropy and Genre

David Shields writes, "As soon as a book can be generically located, it seems to me for all intents and purposes dead . . . When I'm constrained within a form, my mind shuts down, goes on a sitdown strike, saying, 'This is boring, so I refuse to try very hard.'" *Generic* might just be another term for *low-entropy*. In fact, low entropy may be what genre *is*—a kind of prototype or paradigm, a rutted wagon road through the Shannon Game. Roger Ebert observes that when an action hero comes under machine-gun fire, there is a drastically lower

---

14. "A girl on the stairs listens to her father / Beat up her mother," it begins, and ends with what might be a reference to either the mother or the girl: "Someone putting their tongue where their tooth had been."

chance of him coming to harm than, say, if he's attacked by knife. Most viewers subconsciously understand this. Indeed, any piece of art seems to invoke with its inaugural gestures a rather elaborate framework of expectations—by means of which its later gestures tend, on the whole, to be less and less surprising. The mind gradually sits down.

You might have noticed in my Shannon Game attempts that the beginnings of words tend to have higher entropy scores than the latter parts. Matt Mahoney's research at the Florida Institute of Technology has shown that the best text-compression software appears to do better on the second half of a novel than the first. Does this suggest, I wonder, that entropy may be fractal? Do novels and films display the same spike-and-decline pattern that words do?

And for that matter—considering how comparatively bewildered infants are, how comparatively awestruck young children tend to be—does life?

Annie Dillard, in *An American Childhood,* explains her childhood thoughts about literature: "In fact, it was a plain truth that most books fell apart halfway through. They fell apart as their protagonists quit, without any apparent reluctance, like idiots diving voluntarily into buckets, the most interesting part of their lives, and entered upon decades of unrelieved tedium. I was forewarned, and would not so bobble my adult life; when things got dull, I would go to sea."

I think our fairy tales prepare our children for this kind of existential panic about growing up. Nothing is more dispiriting than "And they all lived happily ever after," which means, in information entropy terms, "And then nothing interesting or noteworthy ever happened to them again for the rest of their lives." Or at the very least, "And then you can pretty much imagine what their forties, fifties, and sixties were like, blah, blah, blah, the end." I don't think it would be going too far to argue that these fairy tales sow the seeds of divorce. No one knows what to do after the wedding! Like an entrepreneur who assumed his company would have been bought by now, like an actor out of lines but aware that the cameras are still rolling . . . marriage, for people raised on Western fairy tales, has that same kind of eerie

"Um . . . now what?" quality. "We just, ah, keep on being married, I guess?"

"No one ever asks, 'How did you two stay together?' Everyone always asks, 'How did you two meet?' " a husband, Eric Hayot, laments on an episode of NPR's *This American Life*. The answer to how they stayed together, Hayot has explained, "is the story of like struggle, and, pain, sort of passed through and fought through and overcome. And that's—that's a story you don't tell in public." Nor, it would seem, do you ask about it; even this very segment, ending on these very words, focuses on how he and his wife met. How will we learn?

As for art, the rare work that manages to keep up its entropy for its entire duration can be electrifying. Krzysztof Kieślowski's *Three Colors: White* is a great example of a generically un-locatable film: it's part comedy, part tragedy, part political movie, part detective story, part romance, part anti-romance. At no point do you sense the shape of what's to come. This is the subtlest sort of radicalism—not to push or break the envelope, necessarily, but to force a sort of three-card monte where one never becomes sure which envelope one's in.[15]

Douglas Hofstadter muses in *Gödel, Escher, Bach,* "Perhaps works of art are trying to convey their style more than anything else." I think that when we're reading a book or watching a film, we wonder maybe not so much "Will our hero be rescued?" as "Is this the kind of story where our hero will be rescued?" Perhaps we're interested not so much in the future—*what will happen, what letter comes*

---

15. David Bellos, director of the Program in Translation and Intercultural Communication at Princeton, speculates that firmly "generic" books may be easier for computers to *translate:* "If you were to take a decidedly jaundiced view of some genre of contemporary foreign fiction (say, French novels of adultery and inheritance), you could surmise that since such works have nothing new to say and employ only repeated formulas, then after a sufficient number of translated novels of that kind and their originals had been scanned and put up on the web, Google Translate should be able to do a pretty good simulation of translating other regurgitations of the same ilk . . . For works that are truly original—and therefore worth translating—statistical machine translation hasn't got a hope."

next—as in the present (perfect progressive): *what has been happening, what word have I been spelling.*

Excerpt

Movie previews—I love watching movie previews. Highest entropy you'll get in the whole night. Each clip gives you a whole world.

The way that "ragged claws" are synecdoche for a crustacean, so are anecdotes synecdoche for a life. Poetry reviewers never hesitate to include quotations, samples, but fiction reviewers seem to prefer plot synopsis as a way to give the reader a lossy "thumbnail" of what to expect from the book. Two different lossy compression strategies, each with its own compression artifacts. Try it yourself, as an experiment: try a week of saying to your friends, "Tell me what you did this week," and then a week of saying, "Tell me a story of something that happened to you this week." Summary or excerpt: experiment with which lossy methods work better.

Entropy isn't all about such emotionally detached things as hard-drive space and bandwidth. Data transfer is communication. Surprisal is experience. In the near-paradoxical space between the *size* and *capacity* of a hard disk lies information entropy; in the space between the *size* and *capacity* of a lifetime lies your life.

The Entropy of Counsel

Entropy suggests that we gain the most insight on a question when we take it to the friend, colleague, or mentor of whose reaction and response we're *least certain*.

The Entropy of the Interview

And it suggests, perhaps, reversing the equation, that if we want to gain the most insight into a *person*, we should ask the *question* of whose answer we're least certain.

I remember watching *Oprah* on September 11, 2007; her guests were a group of children who had each lost a parent on September 11, 2001:

> OPRAH: I'm really happy that you all could join us at this time of remembrance. Does it ever get easier—can I ask anybody? Does it ever—

She asks, but the question contains its own response. (Who would dare volunteer, "Yeah, maybe a little," or, "Very, very gradually"?) The question itself creates a kind of moral norm, suggesting—despite evidence to the contrary, in fact[16]—that for a normal person the grief could not have diminished. The coin she's flipping feels two-headed. I grew agitated as the interview went on:

> OPRAH: Do you feel like children of 9/11? Do you feel like that? Do you feel like when somebody knows, Shalisha, that you lost a loved one, that you now have suddenly become a 9/11 kid?
> SHALISHA: I do. I do believe that.
> OPRAH: Well, you know, I said and I've said many times on my show over the years, there isn't a day that goes by that I don't, at some point, think about what happened that day, although I didn't lose anybody that I knew. And opening this show, I said, you all live with it every day. It never goes away, does it?
> AYLEEN: No.

What else can you possibly say to a question like that? First of all, I sincerely doubt that Oprah literally thought about the September 11

---

16. See, e.g., Columbia University clinical psychologist George Bonanno's "Loss, Trauma, and Human Resilience: Have We Underestimated the Human Capacity to Thrive After Extremely Aversive Events?"

attacks every day for six years. Second, how do you expect someone to give an honest answer when you've prefaced the question like that? The guests are being *told* what they feel, not asked.

> OPRAH: And isn't it harder this time of the year?
> KIRSTEN: It's more difficult around this time, I think.

Disappointed, I clicked away. You could practically edit the children's responses out and have the same interview.

Truth be told, I know from reading the transcripts that Oprah's questions do become a little more flexible, and the children do start to open up (the transcripts alone choked me up), but it frustrated me, as a viewer, to see her setting up such rigid containers for their responses. I want to withhold judgment in this particular case: maybe it was a way to ease a group of young, grieving, nervous guests into the conversation—maybe that's even the best tactic for that kind of interview. But on the other hand, or at least in another *situation*, it could come off as an unwillingness to really get to know a person—asking precisely that to which one is most confident of the answer. As a viewer, I felt as if my ability to understand these children was being held back by the questions—as was, I thought, Oprah's. Did she even *want* to know what the kids really felt?

When we think *interview*, we think of a formalized situation, a kind of assessment or sizing up. But etymologically the word means *reciprocal seeing*. And isn't that the aim of all meaningful conversation?

I remember registering a shock upon hitting the passage in *Zen and the Art of Motorcycle Maintenance* where Robert Pirsig says, " 'What's new?' is an interesting and broadening eternal question, but one which, if pursued exclusively, results only in an endless parade of trivia and fashion, the silt of tomorrow. I would like, instead, to be concerned with the question 'What is best?,' a question which cuts deeply rather than broadly, a question whose answers tend to move the silt downstream." I realized: even the basic patterns of conversation can be interrogated. And they can be improved. Information entropy gives us one way in.

Just a few months ago I fell into this trap; recalling the Pirsig quotation got me out. I was detachedly roaming the Internet, but there was nothing interesting happening in the news, nothing interesting happening on Facebook . . . I grew despondent, depressed—the world used to seem so interesting . . . But all of a sudden it dawned on me, as if the thought had just occurred to me, that much of what is interesting and amazing about the world did *not* happen in the past twenty-four hours. How had this fact slipped away from me? (Goethe: "He who cannot draw on three thousand years is living hand to mouth.") Somehow I think the Internet is making this very critical point lost on an entire demographic. Anyway, I read some Thoreau and some Keats and was much happier.

Ditto for the personal sphere. Don't make the mistake of thinking that when "So, what else is new?" runs out of steam you're fully "caught up" with someone. Most of what you don't know about them has little if anything to do with the period between this conversation and your previous one.

Whether in speed dating, political debate, a phone call home, or dinner table conversation, I think information entropy applies. Questions as wide and flat open as the uniform distribution. We learn someone through little surprises. We can learn to talk in a way that elicits them.

Pleasantries are low entropy, biased so far that they stop being an earnest inquiry and become ritual. Ritual has its virtues, of course, and I don't quibble with them in the slightest. But if we really want to start fathoming someone, we need to get them speaking in sentences we can't finish.[17]

---

17. As a confederate, it was often the moments when I (felt I) knew what the judge was typing that I jumped the Q&A gun. This suggests a way in which Shannon Game entropy and the (much less well understood) science

Lempel-Ziv; the Infant Brain; Redefining "Word"

In many compression procedures—most famously, one called the Lempel-Ziv algorithm—bits that occur together frequently get chunked together into single units, which are called *words*. There might be more to that label than it would seem.

It's widely held by contemporary cognitive scientists that infants learn the words of their native language by intuiting which sounds tend, statistically, to occur together most often. I mentioned earlier that Shannon Game values tend to be highest at the starts of words, and lower at the ends: meaning that *intra*-word letter or syllable pairs have significantly lower entropy than *inter*-word pairs. This pattern may be infants' first toehold on English, what enables them to start chunking their parents' sound streams into discrete segments—*words*—that can be manipulated independently. Infants are hip to information entropy before they're hip to their own names. In fact, it's the very thing that gets them there. Remember that oral speech has no pauses or gaps in it—looking at a sound-pressure diagram of speech for the first time, I was shocked to see no inter-word silences—and for much of human history neither did writing. (The space was apparently introduced in the seventh century for the benefit of medieval Irish monks not quite up to snuff on their Latin.) This Shannon entropy spike-and-decay pattern (incidentally, this is also what a musical note looks like on a spectrograph), this downward sloping ramp, may be closer to the root of what a word is than anything having to do with the spacebar.[18]

---

of barge-in may be related: a link between the questions of *how* to finish another's sentences, and *when*.

18. "You know, if people spoke completely compressed text, no one would actually be able to learn English," notes Brown University professor of computer science and cognitive science Eugene Charniak. Likewise, adults would find it much harder to distinguish gibberish at a glance, because every string

And we see the Lempel-Ziv chunking process not just in language *acquisition* but in language *evolution* as well. From "bullpen" to "breadbox" to "spacebar" to "motherfucker," pairings that occur frequently enough fuse into single words.[19] ("In general, permanent compounds begin as temporary compounds that become used so frequently they become established as permanent compounds. Likewise many solid compounds begin as separate words, evolve into hyphenated compounds, and later become solid compounds."[20]) And even when the fusion isn't powerful enough to close the spacebar gap between the two words, or even to solder it with a hyphen, it can often be powerful enough to render the phrase impervious to the

---

of letters or sounds would have at least *some* meaning. "Colorless green ideas sleep furiously" is, famously, nonsensical, but requires a second's thought to identify it as such, whereas "Meck pren plaphth" is gibberish at a glance. A language that was compressed for maximum terseness and economy wouldn't have this distinction.

Another casualty of an optimally compressed language (if anyone could learn it in the first place) would be *crossword puzzles*. As Claude Shannon noted, if our language was better compressed—that is, if words were shorter, with almost all short strings of letters, like "meck" and "pren" and all sorts of others, being valid words—then it would be much harder to complete crossword puzzles, because wrong answers wouldn't produce sections where no words seemed to fit, signaling the error. Intriguingly, with a *less* well-compressed language, with more non-word letter strings and longer words on average, crossword puzzles would be nearly impossible to *compose*, because you couldn't find enough valid words whose spellings crisscrossed in the right way. The entropy of English is just about perfect for crossword puzzles.

19. This sentence read best when I made my examples all nouns, but lest you think that this process happens only to (somewhat uncommon) nouns, and not to *everyday* adjectives and adverbs, I hope you don't think so *anymore*. *Anything* and *everything* can do it.

20. *The American Heritage Book of English Usage,* §8. (That § symbol, being outside both alphabet and punctuation, is probably the entropy value of half a sentence. I get a satisfying feeling from using it. Ditto for other arcane and wonderfully dubbed marks like the pipe, voided lozenge, pilcrow, asterism, and double dagger.)

twists of grammar. Certain phrases imported into English from the Norman French, for example, have stuck so closely together that their inverted syntax never ironed out: "attorney general," "body politic," "court martial." It would seem that these phrases, owing to the frequency of their use, simply came to be taken, subliminally, as atomic, as—internal space be damned!—single words.

So, language learning works like Lempel-Ziv; language evolution works like Lempel-Ziv—what to make of this strange analogue? I put the question to Brown University cognitive scientist Eugene Charniak: "Oh, it's much stronger than just an *analogue*. It's probably what's *actually going on*."

The Shannon Game vs. Your Thumbs: The Hegemony of T9

I'm guessing that if you've ever used a phone to write words—and that is ever closer to being all of us now[21]—you've run up against information entropy. Note how the phone keeps trying to predict what you're saying, what you'll say next. Sound familiar? It's the Shannon Game.

So we have an empirical measure, if we wanted one, of entropy (and maybe, by extension, "literary" value): how often you disappoint your phone. How long it takes you to write. The longer, arguably, and the more frustrating, the more interesting the message might be.

As much as I rely on predictive text capabilities—sending an average of fifty iPhone texts a month, and now even taking down writing ideas on it[22]—I also see them as dangerous: information entropy turned hegemonic. Why hegemonic? Because every time you type a word that isn't the predicted word, you have to (at least

---

21. The most recent statistics I've seen put global cell phone subscriptions at 4.6 billion, in a global population of 6.8 billion.
22. Dave Matthews Band's "You and Me" is, to my knowledge, the first major radio single to have its lyrics written on an iPhone—suggesting that text prediction may increasingly affect not only interpersonal communication but the production of art.

on the iPhone) *explicitly* reject their suggestion or else it's (automatically) substituted. Most of the time this happens, I'm grateful: it smoothes out typos made by mis-hitting the keyboard, which allows for incredibly rapid, reckless texting. But there's the sinister underbelly—and this was just as true too on my previous phone, a standard numerical keypad phone with the T9 prediction algorithm on it. You're gently and sometimes less-than-gently pushed, nudged, bumped into using the language the way the original test group did. (This is particularly true when the algorithm doesn't adapt to your behavior, and many of them, especially the older ones, don't.) As a result, you start unconsciously changing your lexicon to match the words closest to hand. Like the surreal word market in Norton Juster's *Phantom Tollbooth,* certain words become too dear, too pricey, too scarce. That's crazy. That's no way to treat a language. When I type on my laptop keyboard into my word processor, no such text prediction takes place, so my typos don't fix themselves, and I have to type the whole word to say what I intend, not just the start. But I can write what I want. Perhaps I have to type more keystrokes on the average than if I were using text prediction, but there's no disincentive standing between me and the language's more uncommon possibilities. It's worth it.

Carnegie Mellon computer scientist Guy Blelloch suggests the following:

> One might think that lossy text compression would be unacceptable because they are imagining missing or switched characters. Consider instead a system that reworded sentences into a more standard form, or replaced words with synonyms so that the file can be better compressed. Technically the compression would be lossy since the text has changed, but the "meaning" and clarity of the message might be fully maintained, or even improved.

But—Frost—"poetry is what gets lost in translation." And—doesn't it seem—what gets lost in compression?

Establishing "standard" and "nonstandard" ways of using a language necessarily involves some degree of browbeating. (David Foster Wallace's excellent essay "Authority and American Usage" shows how this plays out in dictionary publishing.) I think that "standard" English—along with its subregions of conformity: "academic English," specific fields' and journals' style rules, and so on—has always been a matter of half clarity, half shibboleth. (That "standard" English is not the modally spoken version should be enough to argue for its non-standardness, enough to argue that there is *some* hegemonic force at work, even if unwittingly or benevolently.)

But often *within* communities of speakers and writers, these deviations have gone unnoticed, let alone unpunished: if everyone around you says "ain't," then the idea that "ain't" ain't a word seems ridiculous, *and correctly so.* The modern, globalized world is changing that, however. If American English dominates the Internet, and British-originating searches return mostly American-originating results, then all of a sudden British youths are faced with a daily assault of *u*-less *colors* and *flavors* and *neighbors* like no other generation of Brits before them. Also, consider Microsoft Word: some person or group at Microsoft decided at some point in time which words were in its dictionary and which were not, subtly imposing their own vocabulary on users worldwide.[23] Never before did, say, Baltimorean stevedores or Houston chemists have to care if their vocabulary got the stamp of approval from Seattle-area software engineers: who cared? Now the vocabulary of one group intervenes in communications between members of other groups, flagging perfectly intelligible and standardized terms as mistakes. That said, on the other hand, as long as you can spell it, you can write it (and subsequently force the dictionary to stop red-underlining it). The software doesn't actually *stop* people from typing what they want.

---

23. Including underneath the word "ain't," despite its having been in steady use since the eighteenth century. It returns 83,800,000 results on Google and was said in the 2008 vice presidential debate.

That is, as long as those people are using computers, not phones. Once we're talking about mobile phones, where text prediction schemes rule, things get scarier. In some cases it may be literally impossible to write words the phone doesn't have in its library.

Compression, as noted above, relies on bias—because making expected patterns easier to represent necessarily makes unexpected patterns harder to represent. The yay-for-the-consumer ease of "normal" language use also means there's a penalty for going outside those lines. (For a typewriter-written poem not to capitalize the beginnings of lines, the beginnings of sentences, or the word "I" may be either a sign of laziness *or* an active aesthetic stand taken by the author—but for users subject to auto-"correction," it can only be the latter.)

The more helpful our phones get, the harder it is to be ourselves. For everyone out there fighting to write idiosyncratic, high-entropy, unpredictable, unruly text, swimming upstream of spell-check and predictive auto-completion: *Don't let them banalize you. Keep fighting.*[24]

Compression and the Concept of Time

Systems which involve large amounts of data that go through relatively small changes—a version control system, handling successive versions of a document, or a video compressor, handling successive frames of a film—lend themselves to something called "delta compression." In delta compression, instead of storing a new copy of the data each time, the compressor stores only the original, along with files of the successive changes. These files are referred to as "deltas" or "diffs." Video compression has its own sub-jargon: delta compression goes by "motion compensation," fully stored frames are "key frames" or "I-frames" (intra-coded frames), and the diffs are called "P-frames" (predictive frames).

---

24. In October 2008, an online petition over twenty thousand strong helped persuade Apple to allow users, once they download the new version of the iPhone firmware, to disable auto-correction if they wanted.

The idea, in video compression, is that most frames bear some marked resemblance to the previous frame—say, the lead actor's mouth and eyebrow have moved very slightly, but the static background is exactly the same—thus instead of encoding the entire picture (as with the I-frames), you just (with the P-frames) encode the *diffs* between the last frame and the new one. When the entire scene cuts, you might as well use a new I-frame, because it bears no resemblance to the last frame, so encoding all the diffs will take as long as or longer than just encoding the new image itself. Camera edits tend to contain the same spike and decay of entropy that words do in the Shannon Game.

As with most compression, lowered redundancy means increased fragility: if the original, initial document or key frame is damaged, the diffs become almost worthless and all is lost. In general, errors or noise tends to stick around longer. Also, it's much harder to jump into the middle of a video that's using motion compensation, because in order to render the frame you're jumping to, the decoder must wheel around and look backward for the most recent key frame, prepare that, and then make all of the changes between that frame and the one you want. Indeed, if you've ever wondered what makes streamed online video behave so cantankerously when you try to jump ahead in it, this is a big part of the answer.[25]

But would it be going too far to suggest that delta compression is changing our very understanding of *time*? The frames of a film, each bumped downward by the next; the frames of a View-Master

---

25. Some artists are actually *using* compression artifacts and compression glitches to create a deliberate visual aesthetic, called "datamoshing." From art-world short films like Takeshi Murata's "Monster Movie" to mainstream music videos like the Nabil Elderkin–directed video for Kanye West's "Welcome to Heartbreak," we're seeing a fascinating burst of experiments with what might be called "delta compression mischief." For instance, what happens when you apply a series of diffs to the *wrong* I-frame, and the wall of a subway station starts to furrow and open uncannily, as though it were Kanye West's mouth?

reel, each bumped leftward by the next . . . but these metaphors for motion—each instant in time knocked out of the present by its successor, like bullet casings kicked out of the chamber of an automatic weapon—don't apply to *compressed* video. Time no longer *passes*. The future, rather than displacing it, *revises* the present, spackles over it, touches it up. The past is not the shot-sideways belt of spent moments but the blurry underlayers of the palimpsest, the hues buried in overpainting, the ancient Rome underfoot of the contemporary one. Thought of in this way, a video seems to heap *upward*, one infinitesimally thin layer at a time, toward the eye.

Diffs and Marketing, Personhood

A movie poster gives you one still out of the 172,800-ish frames that make up a feature film, a billboard distills the experience of a week in the Bahamas to a single word, a blurb tries to spear the dozen hours it will take to read a new novel using a trident of just three adjectives. Marketing may be lossy compression pushed to the breaking point. It can teach us things about grammar, by cutting the sentence down to its key word. But if we look specifically at the way art is marketed, we see a pattern very similar to I-frames and P-frames; but in this case, it's a cliché and a diff. Or a genre and a diff.

When artists participate in a stylistic and/or narrative tradition (which is always), we can—and often do—describe their achievement as a diff. Your typical love story, with a twist: _____. Or, just like the sound of _____ but with a note of _____. Or, _____ meets _____.

Children become diffs of their parents. Loves, diffs of old loves. Aesthetics become diffs of aesthetics. This instant: a diff of the one just gone.

Kundera:

> What is unique about the "I" hides itself exactly in what is
> unimaginable about a person. All we are able to imagine is

what makes everyone like everyone else, what people have in common. The individual "I" is what differs from the common stock, that is, what cannot be guessed at or calculated, what must be unveiled.

When the diff between two frames of video is too large (this often occurs across edits or cuts), it's often easier to build a new I-frame than to enumerate all the differences. The analogy to human experience is the moment when, giving up on the insufficient "It's like ____ meets ____" mode of explanation-by-diff, we say "It'd take me longer to explain it to you than to just show you it" or "I can't explain it really, you just have to see it." This, perhaps, as a definition of the sublime?

Diffs and Morality

Thomas Jefferson owned slaves; Aristotle was sexist. Yet we consider them wise? Honorable? Enlightened? But to own slaves in a slave-owning society and to be sexist in a sexist society are low-entropy personality traits. In a compressed biography of people, we leave those out. But we also tend on the whole to pass *less judgment* on the low-entropy aspects of someone's personality compared to the high-entropy aspects. The *diffs* between them and their society are, one could argue, by and large wise and honorable. Does this suggest, then, a *moral* dimension to compression?

Putting In and Pulling Out:
The Eros of Entropy

Douglas Hofstadter:

We feel quite comfortable with the idea that a record contains the same information as a piece of music, because of the existence of record players, which can "read" records and convert the groove-patterns into sounds . . . It is natural, then, to think . . . decoding mechanisms . . . simply reveal information which is intrinsically inside the structures, waiting to be "pulled out." This leads to the idea that for each structure, there are certain pieces of information which *can* be pulled out of it, while there are other pieces of information which *cannot* be pulled out of it. But what does this phrase "pull out" really mean? How hard are you allowed to pull? There are cases where by investing sufficient effort, you can pull very recondite pieces of information out of certain structures. In fact, the pulling-out may involve such complicated operations that it makes you feel you are putting in more information than you are pulling out.

The strange, foggy turf between a decoder pulling information out and putting it in, between implication and inference, is a thriving ground for art criticism and literary translation, as well as that interesting compression technique known as *innuendo,* which thrives on the deniability latent in this in-between space. There's a kind of eros in this, too—*I don't know where you (intention) end and I (interpretation) begin*—as the mere act of listening ropes us into duet.

Men and Women: (Merely?) Players

Part of the question of how good, say, a compressed MP3 sounds is how much of the original uncompressed data is preserved; the other part is how good the MP3 *player* (which is usually also the decompressor) is at guessing, interpolating the values that weren't preserved. To talk about the quality of a *file,* we must consider its relationship to the player.

Likewise, any compression contests or competitions in the computer science community require that participants include the size of the decompressor along with their compressed file. Otherwise you get the "jukebox effect"—"Hey, look, I've compressed Mahler's Second Symphony down to just two bytes! The characters 'A7'! Just punch 'em in and listen!" You can see the song hasn't been compressed at all, but simply moved inside the decompressor.

With humans, however, it works a little differently. The size of our decompressor is fixed—about a hundred billion neurons. Namely, it's huge. So we might as well use it. Why read a book with the detachment of a laser scanning an optical disc? When we engage with art, the world, each other, let us mesh all of our gears, let us seek that which takes maximum advantage of the player—that which calls on our full humanity.

I think the reason novels are regarded to have so much more "information" than films is that they outsource the scenic design and the cinematography to the reader. If characters are said to be "eating eggs," we as readers fill in the plate, silverware, table, chairs, skillet, spatula . . . Granted, each reader's spatula may look different, whereas the film pins it down: *this* spatula, this very one. These specifications demand detailed visual data (ergo, the larger file size of video) but frequently don't matter (ergo, the greater *experienced* complexity of the novel).

This, for me, is a powerful argument for the value and potency of *literature* specifically. Movies don't demand as much from the player. Most people know this; at the end of the day you can be too beat to read but not yet too beat to watch television or listen to music. What's less talked about is the fragility of language: when you watch a foreign film with subtitles, notice that *only the words* have been translated; the cinematography and the soundtrack are perfectly "legible" to you. Even without "translation" of any kind, one can still enjoy and to a large extent appreciate foreign songs and films and sculptures. But that culture's books are just so many squiggles: you try to read a novel in Japanese, for instance, and you get virtually nothing out of the experience. All of this points to how, one might say, *personal* language is. Film and music's power comes in large part from its universality; language's doggedly nonuniversal quality points to a different kind of power altogether.

Pursuit of the Unimaginable

Kundera:

> Isn't making love merely an eternal repetition of the same?
> Not at all. There is always the small part that is unimaginable.
> When he saw a woman in her clothes, he could naturally
> imagine more or less what she would look like naked . . . , but
> between the approximation of the idea and the precision of
> reality there was a small gap of the unimaginable, and it was
> this hiatus that gave him no rest. And then, the pursuit of the
> unimaginable does not stop with the revelations of nudity; it
> goes much further: How would she behave while undressing?
> What would she say when he made love to her? How would
> her sighs sound? How would her face distort at the moment
> of orgasm? . . . He was not obsessed with women; he was
> obsessed with what in each of them is unimaginable . . . So it
> was a desire not for pleasure (the pleasure came as an extra, a
> bonus) but for possession of the world.

The pursuit of the unimaginable, the "will to information," as an argument for womanizing? Breadth before depth? Hardly. But a reminder, I think, that a durable love is one that's dynamic, not static; long-*running*, not long-*standing;* a river we step into every day and not twice. We must dare to find new ways to be ourselves, new ways to discover the unimaginable aspects of ourselves and those closest to us.

Our first months of life, we're in a state of perpetual dumbfoundedness. Then, like a film, like a word, things go—though not without exception—from inscrutable to scrutable to familiar to dull. Unless we are vigilant: this tendency, I believe, can be fought.[26] Maybe it's not so much about possession of the world as a kind of understanding of it. A glint of its insane detail and complexity.

The highest ethical calling, it strikes me, is curiosity. The greatest reverence, the greatest rapture, are in it. My parents tell the story that as a child I went through a few months when just about all I did was point to things and shout, "What's it!" "Ta-ble-cloth." "What's it!" "Nap-kin." "What's it!" "Cup-board." "What's it!" "Pea-nut-but-ter." "What's it!" . . . Bless them, they conferred early on and made the decision to answer every single time with as much enthusiasm as they could muster, never to shut down or silence my inquiry no matter how it grated on them. I started a collection of exceptional sticks by the corner of the driveway that soon came to hold *every stick* I found that week. How can I stay so irrepressibly curious? How can we keep the bit rate of our lives up?

Heather McHugh: We don't care how a poet looks; we care how a poet *looks*.

Forrest Gander: "Maybe the best we can do is try to leave ourselves unprotected. To keep brushing off habits, how we see things and what we expect, as they crust around us. Brushing the green flies of *the usual* off the tablecloth. To pay attention."

The Entropy of English

What the Shannon Game—played over a large enough body of texts and by a large enough group of people—allows us to do is actually quantify the information entropy of written English. Compression relies on probability, as we saw with the coin example, and so English

---

26. E.g., Timothy Ferriss: "My learning curve is insanely steep right now. As soon as that plateaus, I'll disappear to Croatia for a few months or do something else." Not all of us can disappear to Croatia at whim, but the Shannon Game suggests, perhaps, that simply asking the right questions might work.

speakers' ability to anticipate the words in a passage correlates to how compressible the text should be.

Most compression schemes use a kind of pattern matching at the binary level: essentially a kind of find and replace, where long strings of digits that recur in a file are swapped out for shorter strings, and then a kind of "dictionary" is maintained that tells the decompressor how and where to swap the long strings back in. The beauty of this approach is that the compressor looks only at the binary—the algorithm works essentially the same way when compressing audio, text, video, still image, and even computer code itself. When English speakers play the Shannon Game, though, something far trickier is happening. Large and sometimes very abstract things—from spelling to grammar to register to genre—start guiding the reader's guesses. The ideal compression algorithm would know that adjectives tend to come before nouns, and that there are patterns that appear frequently in spelling—"*qu*" being a good example, a pairing *so* common many board games put it on a single tile—all of which reduce the entropy of English. And the ideal compressor would know that "pearlescent" and "dudes" almost never pop up in the same sentence.[27] And that one-word sentences, no matter the word, are too curt, tonally, for legal briefs. And maybe even that twenty-first-century prose tends to use shorter sentences than nineteenth-century prose.

So, what, you may be wondering, *is* the entropy of English? Well, if we restrict ourselves to twenty-six uppercase letters plus the space, we get twenty-seven characters, which, *uncompressed,* requires roughly 4.75 bits per character.[28] But, according to Shannon's 1951 paper "Prediction and Entropy of Printed English," the average entropy of a letter as determined by native speakers playing the Shannon Game comes out to somewhere between 0.6 and 1.3 bits. That is to say, on average, a reader can guess the next letter correctly *half* the time. (Or, from the writer's perspective, as Shannon put it: "When we write

---

27. This sentence itself being one, and perhaps the only, exception.
28. ($\log_2 27 \approx 4.75$)

English half of what we write is determined by the structure of the language and half is chosen freely.") That is to say, a letter contains, on average, the same amount of information—1 bit—as a coin flip.

Entropy and the Turing Test

We return to the Shannon Game one last time. Scientists all the way back to Claude Shannon have regarded creating an optimal playing strategy for this game as equivalent to creating an optimal compression method for English. These two challenges are so related that they amount to one and the same thing.

But only now are researchers[29] arguing one step further—that creating an optimal compressor for English is equivalent to another major challenge in the AI world: passing the Turing test.

If a computer could play this game optimally, they say, if a computer could compress English optimally, it'd know enough about the language that it would *know the language*. We'd have to consider it intelligent—in the human sense of the word.

So a computer, to be humanly intelligent, doesn't even need—as in the traditional Turing test—to respond to your sentences: it need only complete them.

Every time you whip out your mobile and start plowing your thumbs into it—"hey dude 7 sounds good see you there"—you're conducting your very own Turing test; you're seeing if computers have finally caught us or not. Remember that every frustration, every "Why does it keep telling people I'm feeling *I'll* today!?" and "Why in the heck does it keep signing off with *Love, Asian*!?" is, for better or worse, a verdict—and the verdict is *not yet, not just yet*. The line itself is still no match for you. And it's still no match for the person at the other end.

---

29. Florida Tech's Matt Mahoney for one, and Brown's Eugene Charniak for another.

11. Conclusion:
The Most Human Human

The Most Human Computer award in 2009 goes to David Levy—the same David Levy whose politically obsessed "Catherine" took the prize in 1997. Levy's an intriguing guy: he was one of the big early figures in the computer chess scene of the 1980s, and was one of the organizers of the Marion Tinsley–Chinook checkers matches that preceded the Kasparov–Deep Blue showdown in the '90s. He's also the author of the recent nonfiction book *Love and Sex with Robots*, to give you an idea of the other sorts of things that are on his mind when he's not competing for the Loebner Prize.

Levy stands up, to applause, accepts the award from Philip Jackson and Hugh Loebner, and makes a short speech about the importance of AI to a bright future, and the importance of the Loebner Prize to AI. I know what's next on the agenda, and my stomach knots despite itself in the second of interstitial silence before Philip takes back the microphone. I'm certain that Doug's gotten it; he and the Canadian judge were talking NHL from the third sentence in their conversation.

Ridiculous Canadians and their ice hockey, I'm thinking. Then I'm thinking how ridiculous it is that I'm even allowing myself to get this worked up about some silly award—granted, I flew all the way out here to compete for it. Then I'm thinking how ridiculous it is to fly five thousand miles just to have an hour's worth of instant messaging conversation. Then I'm thinking how maybe it'll be great to be the runner-up; I can obsessively scrutinize the transcripts in the book if I want and seem like an underdog, not a gloater. I can figure out what went wrong. I can come back next year, in Los Angeles, with the home-field cultural advantage, and finally show—

"And the results here show also the identification of the *human* that the judges rated 'most human,' " Philip announces, "which as you can see was 'Confederate 1,' which was Brian Christian."

And he hands me the Most Human Human award.

Rivals; Purgatory

I didn't know what to feel about it, exactly. It seemed strange to treat it as meaningless or trivial: I had, after all, prepared quite seriously, and that preparation had, I thought, paid off. And I found myself surprisingly invested in the outcome—how I did individually, yes, but also how the four of us did together. Clearly there was *something* to it all.

On the other hand, I felt equal discomfort regarding my new prize as *significant*—a true measure of *me* as a person—a thought that brought with it feelings of both pride ("Why, I *am* an excellent specimen, and it's kind of you to say so!") and guilt: if I *do* treat this award as "meaning something," how do I act around these three people, my only friends for the next few days of the conference, people judged to be *less human* than myself? What kind of dynamic would that create? (Answer: mostly they just teased me.)

Ultimately, I let that particular question drop: Doug, Dave, and Olga were my comrades far more than they were my foes, and together we'd avenged the mistakes of 2008 in dramatic fashion. 2008's confederates had given up a total of five votes to the computers, and almost allowed one to hit Turing's 30 percent mark, making history. But between us, we hadn't permitted a *single* vote to go the machines' way. 2008 was a nail-biter; 2009 was a rout.

At first this felt disappointing, anticlimactic. There were any number of explanations: there were fewer rounds in '09, so there were simply fewer opportunities for deceptions. The strongest program from '08 was Elbot, the handiwork of a company called Artificial Solutions, one of many new businesses leveraging chatbot technology to "allow our clients to offer better customer service at lower cost." After Elbot's victory at the Loebner Prize competition and the publicity that followed, the company seemingly decided to prioritize the Elbot software's more commercial applications; at any rate, it wouldn't be coming to the '09 contest as returning champion. In some ways it would have been more dramatic to have a closer fight.

In another sense, though, the results were quite dramatic indeed. We think of science as an unhaltable, indefatigable advance: the idea that the Macs and PCs for sale next year would be slower, clunkier, heavier, and more expensive than this year's models is laughable. Even in fields where computers were being matched up to a human standard, such as chess, their advance seemed utterly linear—inevitable, even. Maybe that's because humans were already about as good at these things as they ever were and will ever be. Whereas in conversation it seems we are so complacent so much of the time, so smug, and with so much room for improvement—

In an article about the Turing test, Loebner Prize co-founder Robert Epstein writes, "One thing is certain: whereas the confederates in the competition will never get any smarter, the computers will." I agree with the latter, and couldn't disagree more strongly with the former.

Garry Kasparov says, "Athletes often talk about finding motivation in the desire to meet their own challenges and play their own best game, without worrying about their opponents. Though there is some truth to this, I find it a little disingenuous. While everyone has a unique way to get motivated and stay that way, all athletes thrive on competition, and that means beating someone else, not just setting a personal best . . . We all work harder, run faster, when we know someone is right on our heels . . . I too would have been unable to reach my potential without a nemesis like Karpov breathing down my neck and pushing me every step of the way."

Some people imagine the future of computing as a kind of heaven. Rallying behind an idea called the "Singularity," people like Ray Kurzweil (in *The Singularity Is Near*) and his cohort of believers envision a moment when we make machines smarter than ourselves, who make machines smarter than themselves, and so on, and the whole thing accelerates exponentially toward a massive ultra-intelligence that we can barely fathom. This time will become, in their view, a kind of techno-rapture, where humans can upload their consciousnesses onto the Internet and get assumed, if not bodily, then at least mentally, into an eternal, imperishable afterlife in the world of electricity.

Others imagine the future of computing as a kind of hell. Machines black out the sun, level our cities, seal us in hyperbaric chambers, and siphon our body heat forever.

Somehow, even during my Sunday school days, hell always seemed a little bit unbelievable to me, over the top, and heaven, strangely boring. And both far too static. Reincarnation seemed preferable to either. To me the real, in-flux, changeable and changing world seemed far more interesting, not to mention fun. I'm no futurist, but I suppose, if anything, I prefer to think of the long-term future of AI as neither heaven nor hell but a kind of purgatory: the place where the flawed, good-hearted go to be purified—and tested—and to come out better on the other side.

If Defeat

As for the final verdict on the Turing test itself, in 2010, 2011, and thereafter—

If, or when, a computer wins the gold (*solid* gold, remember) Loebner Prize medal, the Loebner Prize will be discontinued forever. When Garry Kasparov defeated Deep Blue, rather convincingly, in their first encounter in '96, he and IBM readily agreed to return the next year for a rematch. When Deep Blue beat Kasparov (rather less convincingly, I might add) in '97, Kasparov proposed another rematch for '98, but IBM would have none of it. They immediately unplugged

Deep Blue, dismantled it, and boxed up the logs they'd promised to make public.[1] Do you get the unsettling image, as I do, of the heavyweight challenger who, himself, rings the round-ending bell?

The implication seems to be that—because technological evolution seems to occur so much faster than biological evolution, years to millennia—once *Homo sapiens* is overtaken, it won't be able to catch up. Simply put, *the Turing test, once passed, is passed forever.* Frankly, I don't buy it.

IBM's odd anxiousness to basically get out of Dodge after the '97 match suggests a kind of insecurity on their part that I think is very much to the point. The fact is, the human race got to rule the earth—okay, technically, bacteria rule the earth, if you look at biomass, and population, and habitat diversity, but we'll humor ourselves—the fact is, the human race got to where it is by being the most adaptive, flexible, innovative, and quick-learning species on the planet. We're not going to take defeat lying down.

No, I think that, while certainly the first year that computers pass the Turing test will be a historic, epochal one, it does not mark the end of the story. No, I think, indeed, that the *next* year's Turing test will truly be the one to watch—the one where we humans, knocked to the proverbial canvas, must pull ourselves up; the one where we learn how to be *better* friends, artists, teachers, parents, lovers; the one where we *come back.* More human than ever. I want to be there for *that.*

If Victory

And if not defeat, but further rout upon rout? I turn a last time to Kasparov. "Success is the enemy of future success," he says. "One of the most dangerous enemies you can face is complacency. I've

---

1. These logs *would,* three years later, be put on the IBM website, albeit in incomplete form and with so little fanfare that Kasparov himself wouldn't find out about them until 2005.

seen—both in myself and my competitors—how satisfaction can lead to a lack of vigilance, then to mistakes and missed opportunities . . . Winning can convince you everything is fine even if you are on the brink of disaster . . . In the real world, the moment you believe you are entitled to something is exactly when you are ripe to lose it to someone who is fighting harder."

If there's one thing I think the human race has been guilty of for a long time—since antiquity at least—it's a kind of complacency, a kind of entitlement. This is why, for instance, I find it oddly invigorating to catch a cold, come down from my high horse of believing myself a member of evolution's crowning achievement, and get whupped for a couple days by a single-celled organism.

A loss, and the reality check to follow, might do us a world of good.

Maybe the Most Human Human award isn't one that breeds complacency. An "anti-method" doesn't scale, so it can't be "phoned in." And a philosophy of site-specificity means that every new conversation, with every person, in every situation, is a new opportunity to succeed in a unique way—or to fail. Site-specificity doesn't provide the kinds of laurels one can rest on.

It doesn't matter whom you've talked to in the past, how much or how little that dialogue sparkled, what kudos or criticism, if any at all, you got for it.

I walk out of the Brighton Centre, to the bracing sea air for a minute, and into a small, locally owned shoe store looking for a gift to bring back home to my girlfriend; the shopkeeper notices my accent; I tell her I'm from Seattle; she is a grunge fan; I comment on the music playing in the store; she says it's Florence + the Machine; I tell her I like it and that she would probably like Feist . . .

I walk into a tea and scone store called the Mock Turtle and order the British equivalent of coffee and a donut, except it comes with thirteen pieces of dinnerware and nine pieces of flatware; I am *so* in England, I think; an old man, probably in his eighties, is shakily eating a pastry the likes of which I've never seen; I ask him what it is; "coffee meringue," he says and remarks on my accent; an hour later he is telling me about World War II, the exponentially increasing racial diversity of Britain, that *House of Cards* is a pretty accurate depiction of British politics, minus the murders, but that really I should watch *Spooks;* do you get *Spooks* on cable, he is asking me . . .

I meet my old boss for dinner; and after a couple years of being his research assistant and occasionally co-author, and after a brief thought of becoming one of his Ph.D. students, after a year of our paths not really crossing, we negotiate whether our formerly collegial and hierarchical relationship, now that its context is removed, simply dries up or flourishes into a domain-general friendship; we are ordering appetizers and saying something about Wikipedia, something about Thomas Bayes, something about vegetarian dining . . .

Laurels are of no use. If you de-anonymized yourself in the past, great. But that was that. And now, you begin again.

Epilogue: The Unsung Beauty
of the Glassware Cabinet

The Most Room-Like Room: The Cornell Box

The image-processing world, it turns out, has a close analogue to the Turing test, called "the Cornell box," which is a small model of a room with one red wall and one green wall (the others are white) and two blocks sitting inside it. Developed by Cornell University graphics researchers in 1984, the box has evolved and become more sophisticated over time, as researchers attempt additional effects (reflection, refraction, and so on). The basic idea is that the researchers set up this room in real life, photograph it, and put the photographs online; graphics teams, naturally, try to get their *virtual* Cornell box renderings to look as much as possible like the real thing.

Of course this raises some great questions.

Graphics teams don't use the Cornell box as a *competitive* standard, and there's an assumption of good faith on their part when they show off their renderings. Obviously, one could simply scan the real photograph and have software output the image, pixel for pixel. As with the Turing test, a static demo won't do. One needs some degree of "interaction" between the judges and the software—in this case, something like moving some of the internal boxes around, or changing the colors, or making one of the boxes reflective, and so on.

Second is that if this particular room is meant to stand in for *all* of visual reality—the way a Turing test is meant to stand in for all of language use—then we might ask certain questions about the room. What kind of light is trickiest? What types of surfaces are the hardest to virtualize? How, that is, do we get the real Cornell box to be a good confederate, the Most Room-Like Room?

My friend Devon does computer-generated imagery (CGI) for animated feature films. The world of CGI movies is a funny place; it takes its cues from reality, yet its aim is not necessarily realism. (Though, he notes, "your range of what's believable is wider than reality.")

Being a computer graphics person brings with it, as most jobs do, a certain way of looking at and of noticing the world. My own poetry background, for instance, gives me an urge to read things against the grain of the author's intended meaning. I read a newspaper headline the other day that said, "UK Minister's Charm Offensive." This to me was hilarious. Of course they meant "offensive" as a noun, as in the tactical deployment of charm for diplomacy purposes, but I kept reading it as an adjective, as though the minister's creepy unctuousness had really crossed the line this time. My friends in the police force and the military can't enter a room without sussing out its entrances and exits; for the one in the fire department, it's alarms and extinguishers.

But Devon: What does a computer graphics guy look for?

"Sharp edges—if you're looking at, like, anything, any sort of man-made object, if it has sharp edges, like a building, or a table: if all the edges are really sharp, then that's a pretty good sign. If you look in the corners of a lit room—if the corners aren't appropriately dark, or too dark . . . Just like complexity of surfaces and irregularities—any type of irregularity, you know. That's all totally—if it's computer generated—it's really hard to do. You look for the quantity of irregularities and regularities, even textures, and things like that. But that's all pretty basic stuff. At another level, you have to start thinking about, like, light bouncing off things, you know, like if you have, for instance, a red wall next to a white wall, how much of the red gets onto the white, and that's the sort of thing that can sort of throw you off."

Of course, as he's saying this to me on the phone, I'm looking around the room, and I'm noticing, as if for the first time, the weird ways that light and shadow seem to bunch up in the corners and along the edges—authentically, I *guess*—I look out the window at the sky—and how many times have you looked at a sky and said, "If this were in a movie, I would criticize the special effects"?

> *Should you paint*
> *a credible sky*
> *you must keep in mind*
> *its essential phoniness.*

—EDUARDO HURTADO

Devon's most recent assignment had been to work on rocket-launcher contrails, a problem that proved trickier than he'd originally thought; he stayed late many long evenings trying to get its waviness and dispersion just *so*. He finally nailed it, and the studio was pleased: it went into the film. But all that scrutiny came with a price. Now, when he goes outside and looks at airplane contrails, he's *suspicious*. "When I was working on those smoke trail things—when I was out hiking or something, watching the planes go by, and trying to analyze how the shape changed over time . . . You almost, like, question reality at times—like, you're looking at something, like smoke or something, and you think, that's too regular, the smoke shouldn't look so regular . . ."

That's what I keep feeling, now, when I read an email or pick up the phone. Even with my own parents—I found myself waiting, like the phonagnosic Steve Royster, for the moment they said something incontrovertibly, inimitably "them."

The Unsung Beauty of the Glassware Cabinet

Curious learning not only makes unpleasant things
less unpleasant, but also makes pleasant things more
pleasant. I have enjoyed peaches and apricots more
since I have known that they were first cultivated in
China in the early days of the Han dynasty; that Chi-
nese hostages held by the great King Kaniska intro-
duced them into India, whence they spread to Persia,
reaching the Roman Empire in the first century of
our era; that the word "apricot" is derived from the
same Latin source as the word "precocious," because
the apricot ripens early; and that the A at the begin-
ning was added by mistake, owing to a false etymol-
ogy. All this makes the fruit taste much sweeter.

—BERTRAND RUSSELL

Reflection and refraction are difficult to simulate on a computer. So is water distortion. So-called "caustics," the way that a glass of wine refocuses its light into a red point on your table, are particularly hard to render.

Reflection and refraction are also fairly computationally nasty because they have the habit of multiplying off of each other. You put two mirrors in front of each other, and the images multiply to infinity in no time flat. Light travels roughly 200,000 miles per second: that's a lot of ping-pong, and way beyond the point where most rendering algorithms tap out. Usually a programmer will specify the maximum acceptable number of reflections or refractions and cap it there, after which point a kind of software deus ex machina sends the light directly back to the eye: no more bouncing.

Getting off the phone with Devon, I go to my kitchen and fling open the glassware cabinet. I am more mesmerized by the hall of mirrors within than I have ever been before. My eyeball bulging at the side of a wineglass, I watch real life, real physics, real light *per-form*.

A glassware cabinet is a computational nightmare.

So, Devon explains, is a deciduous forest. And nude bodies are more of a computational nightmare than clothed ones: all those tiny hairs, irregular curvatures, semi-translucencies of veins under slightly mottled skin.

I love these moments when the theory, the models, the approximations, as good as they are, aren't good enough. You simply must watch. *Ah, so* this *is how nature does it.* This *is what it looks like.* I think it's important to know these things, to know what can't be simulated, can't be made up, can't be imagined—and to seek it.

Devon, in his life out of the studio, now pays a kind of religious attention to the natural world. It helps him do a better job in his animation projects, I'm sure, but one suspects the means and ends are actually the other way around.

"It's nice to know at least that there are quite a few things that, at least with computer graphics, and what I do, that I'm like, I mean, *Wow.* You know, I wrote some *thing,* and I have people waiting, you know, ten hours for a *frame,* and it doesn't even look realistic, it doesn't even look quite right! And I'm like, *Damn*—that's, one, quite far from reality, and, two, it's stretching, like, x number of dollars' worth of computing at it and barely even making it. *So.*"

Devon laughs.

"It feels . . . It definitely feels good at the end of the day that I can open my eyes and look at something that's, like, many orders of magnitude more complex."

And to be able to know *where* to look for it—

And how to recognize it.

Acknowledgments

It was Isaac Newton who famously said (though it was actually a common expression at the time), "If I have seen a little further it is by standing on the shoulders of giants." I want to say, more neurologically, that if I've been able to conduct a good signal to my axon terminal, I owe it to the people at my dendrites. (Though it goes without saying, of course, that any noise or error in the signal is my own.)

I'm indebted to a number of conversations with friends and colleagues, which sparked or contributed many of the specific ideas in the text. I recall, in particular, such conversations with Richard Kenney, David Shields, Tom Griffiths, Sarah Greenleaf, Graff Haley, François Briand, Greg Jensen, Joe Swain, Megan Groth, Matt Richards, Emily Pudalov, Hillary Dixler, Brittany Dennison, Lee Gilman, Jessica Day, Sameer Shariff, Lindsey Baggette, Alex Walton, Eric Eagle, James Rutherford, Stefanie Simons, Ashley Meyer, Don Creedon, and Devon Penney.

Thanks to the researchers and experts of their respective crafts who graciously volunteered their time to speak at length in person (or the closest technological equivalent): Eugene Charniak, Melissa Prober, Michael Martinez, Stuart Shieber, Dave Ackley, David Sheff, Kevin Warwick, Hava Siegelmann, Bernard Reginster, Hugh Loebner, Philip Jackson, Shalom Lappin, Alan Garnham, John Carroll, Rollo Carpenter, Mohan Embar, Simon Laven, and Erwin van Lun.

Thanks, too, to those with whom I corresponded by email, who offered thoughts and/or pointed me toward important research: Daniel Dennett, Noam Chomsky, Simon Liversedge, Hazel Blythe, Dan Mirman, Jenny Saffran, Larry Grobel, Daniel Swingley, Lina Zhou, Roberto Caminiti, Daniel Gilbert, and Matt Mahoney.

Thanks to the University of Washington Libraries and the Seattle Public Library; I am in your debt, quite literally.

Thanks to Graff Haley, Matt Richards, Catherine Imbriglio, Sarah Greenleaf, Randy Christian, Betsy Christian, and, with special appreciation, Greg Jensen, all of whom read and offered feedback on an earlier draft.

Thanks to Sven Birkerts and Bill Pierce at *AGNI,* for publishing an earlier version of "High Surprisal" (as "High Compression: Information, Intimacy, and the Entropy of Life") in their pages, for their sharp editorial eyes and their support.

Thanks to my agent, Janet Silver at Zachary Shuster Harmsworth, for believing in the project from day one, and for her support, wisdom, and enthusiasm throughout.

Thanks to my editors, Bill Thomas and Melissa Danaczko, and the rest of the Doubleday team, for their expert eyes, and for all of the faith and hard work of bringing the book into the world.

Thanks to invaluable fellowships at the Bread Loaf Writers' Conference in Ripton, Vermont; at Yaddo in Saratoga Springs, New York; and at the MacDowell Colony in Peterborough, New Hampshire. A reverence for good work fills them with a kind of airborne sacredness like very few places I know.

Thanks to the baristas of Capitol Hill and Wallingford for the liquid jumper cables of many Seattle mornings.

Thanks to a hamster-sitting residency at the Osborn/Coleman household, where good work was done.

Thanks to Michael Langan for a very fine portrait.

Thanks to Philip Jackson, for allowing me to be a part of the 2009 Loebner Prize competition, and to my fellow confederates, Dave Marks, Doug Peters, and Olga Martirosian, with whom I was proud to represent humanity.

Thanks to my parents, Randy Christian and Betsy Christian, for the unconditional everything along the way.

Thanks to the inestimable Sarah Greenleaf, whose clarity of mind cut many a Gordian knot, and whose courage and compassion have shaped both the text and its author.

Thanks to everyone who has taught me, by words or by example, what it means to be human.

Notes

Epigraphs

v David Foster Wallace, in interview with David Lipsky, in *Although of Course You End Up Becoming Yourself* (New York: Broadway Books, 2010).

xi Richard Wilbur, "The Beautiful Changes," *The Beautiful Changes and Other Poems* (New York: Reynal & Hitchcock, 1947).

xi Robert Pirsig, *Zen and the Art of Motorcycle Maintenance* (New York: Morrow, 1974).

xi Barack Obama, "Remarks by the President on the 'Education to Innovate' Campaign," press release, The White House, Office of the Press Secretary, November 23, 2009.

0. Prologue

1 See, e.g., Neil J. A. Sloane and Aaron D. Wyner, "Biography of Claude Elwood Shannon," in *Claude Elwood Shannon: Collected Papers* (New York: IEEE Press, 1993).

1. Introduction: The Most Human Human

4 Alan Turing, "Computing Machinery and Intelligence," *Mind* 59, no. 236 (October 1950), pp. 433–60.

4 Turing initially introduces the Turing test by way of analogy to a game in which a judge is conversing over "teleprinter" with two *humans,* a man and a woman, both of whom are claiming to be the woman. Owing to some ambiguity in Turing's phrasing, it's not completely clear how

strong of an analogy he has in mind; for example, is he suggesting that in the Turing test, a woman and a computer are both claiming specifically to be a woman? Some scholars have argued that the scientific community has essentially swept this question of gender under the rug in the subsequent (gender-neutral) history of the Turing test, but in BBC radio interviews in 1951 and 1952, Turing makes it clear (using the word "man," which is gender-neutral in the context) that he is, in fact, talking about a *human* and a machine both claiming to be *human*, and therefore that the gender game was merely an example to help explain the basic premise at first. For an excellent discussion of the above, see Stuart Shieber, ed., *The Turing Test: Verbal Behavior as the Hallmark of Intelligence* (Cambridge, Mass.: MIT Press, 2004).

5 Charles Platt, "What's It Mean to Be Human, Anyway?" *Wired*, no. 3.04 (April 1995).

6 Hugh Loebner's Home Page, www.loebner.net.

6 Hugh Loebner, letter to the editor, *New York Times*, August 18, 1994.

7 *The Terminator*, directed by James Cameron (Orion Pictures, 1984).

7 *The Matrix*, directed by Andy Wachowski and Larry Wachowski (Warner Bros., 1999).

8 *Parsing the Turing Test*, edited by Robert Epstein et al. (New York: Springer, 2008).

8 Robert Epstein, "From Russia, with Love," *Scientific American Mind*, October/November 2007.

9 97 percent of all email messages are spam: Darren Waters, citing a Microsoft security report, in "Spam Overwhelms E-Mail Messages," *BBC News*, April 8, 2009, news.bbc.co.uk/2/hi/technology/7988579.stm.

9 Say, Ireland: Ireland consumes 25,120,000 megawatt hours of electricity annually, according to the CIA's *World Factbook*, www.cia.gov/library/publications/the-world-factbook/rankorder/2042rank.html. The processing of spam email consumes 33,000,000 megawatt hours annually worldwide, according to McAfee, Inc., and ICF International's 2009 study, "The Carbon Footprint of Email Spam Report," newsroom.mcafee.com/images/10039/carbonfootprint2009.pdf.

10 David Alan Grier, *When Computers Were Human* (Princeton, N.J.: Princeton University Press, 2005).

11 Daniel Gilbert, *Stumbling on Happiness* (New York: Knopf, 2006).

11 Michael Gazzaniga, *Human: The Science Behind What Makes Us Unique* (New York: Ecco, 2008).

12 Julian K. Finn, Tom Tregenza, and Mark D. Norman, "Defensive Tool Use in a Coconut-Carrying Octopus," *Current Biology* 19, no. 23 (December 15, 2009), pp. 1069–70.

12 Douglas R. Hofstadter, *Gödel, Escher, Bach: An Eternal Golden Braid* (New York: Basic Books, 1979).

13 Noam Chomsky, email correspondence (emphasis mine).

13 John Lucas, "Commentary on Turing's 'Computing Machinery and Intelligence,' " in Epstein et al., *Parsing the Turing Test*.

2. *Authenticating*

16 Alix Spiegel, " 'Voice Blind' Man Befuddled by Mysterious Callers," *Morning Edition,* National Public Radio, July 12, 2010.

17 David Kernell, posting (under the handle "rubico") to the message board www.4chan.org, September 17, 2008.

18 Donald Barthelme,"Not-Knowing," in *Not-Knowing: The Essays and Interviews of Donald Barthelme,* edited by Kim Herzinger (New York: Random House, 1997). Regarding "Bless Babel": Programmers have a concept called "security through diversity," which is basically the idea that a world with a number of different operating systems, spreadsheet programs, etc., is more secure than one with a software "monoculture." The idea is that the effectiveness of a particular hacking technique is limited to the machines that "speak that language," the way that genetic diversity generally means that no single disease will wipe out an entire species. Modern operating systems are designed to be "idiosyncratic" about how certain critical sections of memory are allocated, so that each computer, even if it is running the same basic environment, will be a little bit different. For more, see, e.g., Elena Gabriela Barrantes, David H. Ackley, Stephanie Forrest, Trek S. Palmer, Darko Stefanovic, and Dino Dai Zovi, "Intrusion Detection: Randomized Instruction Set Emulation to Disrupt Binary Code Injection Attacks," *Proceedings of the 10th ACM Conference on Computer and Communication Security* (New York: ACM, 2003), pp. 281–89.

19 "Speed Dating with Yaacov and Sue Deyo," interview by Terry Gross, *Fresh Air,* National Public Radio, August 17, 2005. See also Yaacov

Deyo and Sue Deyo, *Speed Dating: The Smarter, Faster Way to Lasting Love* (New York: HarperResource, 2002).

19 "Don't Ask, Don't Tell," season 3, episode 12 of *Sex and the City*, August 27, 2000.

20 For more on how the form/content problem in dating intersects with computers, see the excellent video by the Duke University behavioral economist Dan Ariely, "Why Online Dating Is So Unsatisfying," Big Think, July 7, 2010, bigthink.com/ideas/20749.

20 The 1991 Loebner Prize transcripts, unlike most other years, are unavailable through the Loebner Prize website. The Clay transcripts come by way of Mark Halpern, "The Trouble with the Turing Test," *New Atlantis* (Winter 2006). The Weintraub transcripts, and judge's reaction, come by way of P. J. Skerrett, "Whimsical Software Wins a Prize for Humanness," *Popular Science*, May 1992.

25 Rollo Carpenter, personal interview.

25 Rollo Carpenter, in "PopSci's Future of Communication: Cleverbot," Science Channel, October 6, 2009.

26 Bernard Reginster (lecture, Brown University, October 15, 2003).

26 "giving style to one's character": Friedrich Nietzsche, *The Gay Science*, translated by Walter Kaufman (New York: Vintage, 1974), sec. 290.

26 Jaron Lanier, *You Are Not a Gadget: A Manifesto* (New York: Knopf, 2010).

28 Eugene Demchenko and Vladimir Veselov, "Who Fools Whom?" in *Parsing the Turing Test,* edited by Robert Epstein et al. (New York: Springer, 2008).

29 *Say Anything . . . ,* directed and written by Cameron Crowe (20th Century Fox, 1989).

29 Robert Lockhart, "Integrating Semantics and Empirical Language Data" (lecture at the Chatbots 3.0 conference, Philadelphia, March 27, 2010).

30 For more on Google Translate, the United Nations, and literature, see, e.g., David Bellos, "I, Translator," *New York Times,* March 20, 2010; and Miguel Helft, "Google's Computing Power Refines Translation Tool," *New York Times,* March 8, 2010.

30 *The Office,* directed and written by Ricky Gervais and Stephen Merchant, BBC Two, 2001–3.

31 Hilary Stout, "The End of the Best Friend," also titled "A Best Friend? You Must Be Kidding," *New York Times,* June 16, 2010.

34 *50 First Dates,* directed by Peter Segal (Columbia Pictures, 2004).

34 Jennifer E. Whiting, "Impersonal Friends," *Monist* 74 (1991), pp. 3–29. See also Jennifer E. Whiting, "Friends and Future Selves," *Philosophical Review* 95 (1986), pp. 547–80; and Bennett Helm, "Friendship," in *The Stanford Encyclopedia of Philosophy,* edited by Edward N. Zalta (Fall 2009 ed.).

35 Richard S. Wallace, "The Anatomy of A.L.I.C.E.," in Epstein et al., *Parsing the Turing Test.*

36 For more on MGonz, see Mark Humphrys, "How My Program Passed the Turing Test," in Epstein et al., *Parsing the Turing Test.*

3. The Migratory Soul

39 Hiromi Kobayashi and Shiro Kohshima, "Unique Morphology of the Human Eye," *Nature* 387, no. 6635, June 19, 1997, pp. 767–68.

39 Michael Tomasello et al., "Reliance on Head Versus Eyes in the Gaze Following of Great Apes and Human Infants: The Cooperative Eye Hypothesis," *Journal of Human Evolution* 52, no. 3 (March 2007), pp. 314–20.

39 Gert-Jan Lokhorst, "Descartes and the Pineal Gland," in *The Stanford Encyclopedia of Philosophy,* edited by Edward N. Zalta (Spring 2009 ed.).

40 Carl Zimmer, *Soul Made Flesh: The Discovery of the Brain—and How It Changed the World* (New York: Free Press, 2004).

41 *Karšu* and the other terms: Leo G. Perdue, *The Sword and the Stylus: An Introduction to Wisdom in the Age of Empires* (Grand Rapids, Mich.: W. B. Eerdmans, 2008). See also Dale Launderville, *Spirit and Reason: The Embodied Character of Ezekiel's Symbolic Thinking* (Waco, Tex.: Baylor University Press, 2007).

41 "black wires grow on her head": The Shakespeare poem is the famous Sonnet 130, "My mistress' eyes are nothing like the sun . . ."

42 Hendrik Lorenz, "Ancient Theories of Soul," in *The Stanford Encyclopedia of Philosophy,* edited by Edward N. Zalta (Summer 2009 ed.).

42 "A piece of your brain": V. S. Ramachandran and Sandra Blakeslee,

Phantoms in the Brain: Probing the Mysteries of the Human Mind (New York: William Morrow, 1998).

44 *All Dogs Go to Heaven,* directed by Don Bluth (Goldcrest, 1989).

44 *Chocolat,* directed by Lasse Hallström (Miramax, 2000).

46 Friedrich Nietzsche, *The Complete Works of Friedrich Nietzsche, Volume 4: The Will to Power, Book One and Two,* translated by Oscar Levy (London: George Allen and Unwin, 1924), sec. 75.

46 Aristotle, *The Nicomachean Ethics,* translated by J. A. K. Thomson and Hugh Tredennick (London: Penguin, 2004), 1178b5–25.

49 Claude Shannon, "A Symbolic Analysis of Relay and Switching Circuits" (master's thesis, Massachusetts Institute of Technology, 1940).

50 President's Commission for the Study of Ethical Problems in Medicine and Biomedical and Behavioral Research, *Defining Death: Medical, Legal, and Critical Issues in the Determination of Death* (Washington, D.C.: U.S. Government Printing Office, 1981).

50 Ad Hoc Committee of the Harvard Medical School to Examine the Definition of Brain Death, "A Definition of Irreversible Coma," *Journal of the American Medical Association* 205, no. 6 (August 1968), pp. 337–40.

51 The National Conference of Commissioners on Uniform State Laws, Uniform Determination of Death Act (1981).

52 Michael Gazzaniga, "The Split Brain Revisited," *Scientific American* (2002). See also the numerous videos available on YouTube of Gazzaniga's interviews and research: "Early Split Brain Research: Michael Gazzaniga Interview," www.youtube.com/watch?v=0lmfxQ -HK7Y; "Split Brain Behavioral Experiments," www.youtube.com/ watch?v=ZMLzP1VCANo; "Split-Brain Patients," www.youtube.com/ watch?v=MZnyQewsB_Y.

53 "You guys are just so *funny*": Ramachandran and Blakeslee, *Phantoms in the Brain,* citing Itzhak Fried, Charles L. Wilson, Katherine A. MacDonald, and Eric J. Behnke, "Electric Current Stimulates Laughter," *Nature* 391 (February 1998), p. 650.

54 a woman gave her number to male hikers: Donald G. Dutton and Arthur P. Aron, "Some Evidence for Heightened Sexual Attraction Under Conditions of High Anxiety," *Journal of Personality and Social Psychology* 30 (1974).

55 Oliver Sacks, *The Man Who Mistook His Wife for a Hat* (New York: Summit Books, 1985).

56 Ramachandran and Blakeslee, *Phantoms in the Brain*.

56 Ken Robinson, "Ken Robinson Says Schools Kill Creativity," TED.com.

57 Ken Robinson, "Transform Education? Yes, We Must," Huffington Post, January 11, 2009.

58 Baba Shiv, "The Frinky Science of the Human Mind" (lecture, 2009).

58 Dan Ariely, *Predictably Irrational* (New York: Harper, 2008).

58 Dan Ariely, *The Upside of Irrationality: The Unexpected Benefits of Defying Logic at Work and at Home* (New York: Harper, 2010).

59 Daniel Kahneman, "A Short Course in Thinking About Thinking" (lecture series), Edge Master Class 07, Auberge du Soleil, Rutherford, Calif., July 20–22, 2007, www.edge.org/3rd_culture/kahneman07/kahneman07_index.html.

60 Antoine Bechara, "Choice," *Radiolab*, November 14, 2008.

60 *Blade Runner,* directed by Ridley Scott (Warner Bros., 1982).

60 Philip K. Dick, *Do Androids Dream of Electric Sheep?* (Garden City, N.Y.: Doubleday, 1968).

62 William Butler Yeats, "Sailing to Byzantium," in *The Tower* (New York: Macmillan, 1928).

63 Dave Ackley, personal interview.

64 Ray Kurzweil, *The Singularity Is Near: When Humans Transcend Biology* (New York: Viking, 2005).

64 Hava Siegelmann, personal interview.

65 See Jessica Riskin, "The Defecating Duck; or, The Ambiguous Origins of Artificial Life," *Critical Inquiry* 20, no. 4 (Summer 2003), pp. 599–633.

66 Roger Levy, personal interview.

67 Jim Giles, "Google Tops Translation Ranking," *Nature News*, November 7, 2006. See also Bill Softky, "How Google Translates Without Understanding," *The Register,* May 15, 2007; and the official NIST results from 2006 at http://www.itl.nist.gov/iad/mig/tests/mt/2006/doc/mt06eval_official_results.html. It's worth noting that particularly in languages like German with major syntactical discrepancies from English, where a word in a sentence of the source language can appear in a very distant place in the sentence of the target language, a purely

statistical approach is not quite as successful, and some hard-coding (or inference) of actual syntactical *rules* (e.g., "sentences generally have a 'subject' portion and a 'predicate' portion") will indeed help the translation software.

69 Randall C. Kennedy, "Fat, Fatter, Fattest: Microsoft's Kings of Bloat," *InfoWorld*, April 14, 2008.

71 W. Chan Kim and Renée Mauborgne, *Blue Ocean Strategy: How to Create Uncontested Market Space and Make the Competition Irrelevant* (Boston: Harvard Business School Press, 2005).

72 awe: It appears that articles that inspire awe are the most likely to be emailed or become "viral," counter to popular thinking that fear, sex, and/or irony prevail online. See John Tierney, "People Share News Online That Inspires Awe, Researchers Find," *New York Times*, February 8, 2010, which cites the University of Pennsylvania's Jonah Berger and Katherine Milkman's study, "Social Transmission and Viral Culture."

4. Site-Specificity vs. Pure Technique

75 Joseph Weizenbaum, *Computer Power and Human Reason: From Judgment to Calculation* (San Francisco: W. H. Freeman, 1976).

75 Joseph Weizenbaum, "ELIZA—a Computer Program for the Study of Natural Language Communication Between Man and Machine," *Communications of the Association for Computing Machinery* 9, no. 1 (January 1966), pp. 36–45.

75 To be precise, ELIZA was a software framework or paradigm developed by Weizenbaum, who actually wrote a number of different "scripts" for that framework. The most famous of these by far is the Rogerian therapist persona, which was called DOCTOR. However, "ELIZA running the DOCTOR script" is generally what people mean when they refer to "ELIZA," and for brevity and ease of understanding I've followed the convention (used by Weizenbaum himself) of simply saying "ELIZA."

76 Kenneth Mark Colby, James B. Watt, and John P. Gilbert, "A Computer Method of Psychotherapy: Preliminary Communication," *Journal of Nervous and Mental Disease* 142, no. 2 (February 1966).

76 Carl Sagan, in *Natural History* 84, no. 1 (January 1975), p. 10.

76 National Institute for Health and Clinical Excellence, "Depression and Anxiety: Computerised Cognitive Behavioural Therapy (CCBT)," www.nice.org.uk/guidance/TA97.

77 Dennis Greenberger and Christine A. Padesky, *Mind Over Mood: Change How You Feel by Changing the Way You Think* (New York: Guilford, 1995).

77 Sting, "All This Time," *The Soul Cages* (A&M, 1990).

78 Richard Bandler and John Grinder, *Frogs into Princes: Neuro Linguistic Programming* (Moab, Utah: Real People Press, 1979).

79 Weizenbaum, *Computer Power and Human Reason.*

79 Josué Harari and David Bell, introduction to *Hermes,* by Michel Serres (Baltimore: Johns Hopkins University Press, 1982).

80 Jason Fried and David Heinemeier Hansson, *Rework* (New York: Crown Business, 2010).

80 Timothy Ferriss, *The 4-Hour Workweek: Escape 9–5, Live Anywhere, and Join the New Rich* (New York: Crown, 2007).

82 Bill Venners, "Don't Live with Broken Windows: A Conversation with Andy Hunt and Dave Thomas," *Artima Developer,* March 3, 2003, www.artima.com/intv/fixit.html.

83 U.S. Marine Corps, *Warfighting.*

83 "NUMMI," episode 403 of *This American Life,* March 26, 2010.

84 Studs Terkel, *Working: People Talk About What They Do All Day and How They Feel About What They Do* (New York: Pantheon, 1974).

85 Matthew B. Crawford, *Shop Class as Soulcraft: An Inquiry into the Value of Work* (New York: Penguin, 2009).

88 Robert Pirsig, *Zen and the Art of Motorcycle Maintenance* (New York: Morrow, 1974).

89 Francis Ponge, *Selected Poems* (Winston-Salem, N.C.: Wake Forest University Press, 1994).

89 Garry Kasparov, *How Life Imitates Chess* (New York: Bloomsbury, 2007).

90 Twyla Tharp, *The Creative Habit: Learn It and Use It for Life* (New York: Simon & Schuster, 2003).

90 "Australian Architect Becomes the 2002 Laureate of the Pritzker Architecture Prize," *Pritzker Architecture Prize,* www.pritzkerprize .com/laureates/2002/announcement.html.

90 "Life is not about maximizing everything": From Geraldine O'Brien,

"The Aussie Tin Shed Is Now a World-Beater," *Sydney Morning Herald,* April 15, 2002.

90 "One of the great problems of our period": From Andrea Oppenheimer Dean, "Gold Medal: Glenn Murcutt" (interview), *Architectural Record,* May 2009.

91 "I think that one of the disasters": Jean Nouvel, interviewed on *The Charlie Rose Show,* April 15, 2010.

91 "I fight for specific architecture": From Jacob Adelman, "France's Jean Nouvel Wins Pritzker, Highest Honor for Architecture," Associated Press, March 31, 2008.

91 "I try to be a contextual architect": *Charlie Rose,* April 15, 2010.

91 "It's great arrogance": From Belinda Luscombe, "Glenn Murcutt: Staying Cool Is a Breeze," *Time,* August 26, 2002.

92 *My Dinner with Andre,* directed by Louis Malle (Saga, 1981).

92 *Before Sunrise,* directed by Richard Linklater (Castle Rock Entertainment, 1995).

92 Roger Ebert, review of *My Dinner with Andre,* January 1, 1981, at rogerebert.suntimes.com.

94 *Before Sunset,* directed by Richard Linklater (Warner Independent Pictures, 2004).

94 George Orwell, "Politics and the English Language," *Horizon* 13, no. 76 (April 1946), pp. 252–65.

98 Melinda Bargreen, "Violetta: The Ultimate Challenge," interview with Nuccia Focile, in program for Seattle Opera's *La Traviata,* October 2009.

5. *Getting Out of Book*

99 Paul Ekman, *Telling Lies: Clues to Deceit in the Marketplace, Politics, and Marriage* (New York: Norton, 2001).

99 Benjamin Franklin, "The Morals of Chess," *Columbian Magazine* (December 1786).

102 For Deep Blue engineer Feng-hsiung Hsu's take on the match, see *Behind Deep Blue: Building the Computer That Defeated the World Chess Champion* (Princeton, N.J.: Princeton University Press, 2002).

103 Neil Strauss, *The Game: Penetrating the Secret Society of Pickup Artists* (New York: ReganBooks, 2005).

103 Duchamp's quotation is attributed to two separate sources: Andy Soltis, "Duchamp and the Art of Chess Appeal," n.d., unidentified newspaper clipping, object file, Department of Modern and Contemporary Art, Philadelphia Museum of Art; and Marcel Duchamp's address on August 30, 1952, to the New York State Chess Association; see Anne d'Harnoncourt and Kynaston McShine, eds., *Marcel Duchamp* (New York: Museum of Modern Art, 1973), p. 131.

104 Douglas R. Hofstadter, *Gödel, Escher, Bach: An Eternal Golden Braid* (New York: Basic Books, 1979).

104 "the conclusion that profoundly insightful chess-playing": Douglas Hofstadter, summarizing the position taken by *Gödel, Escher, Bach* in the essay "Staring Emmy Straight in the Eye—and Doing My Best Not to Flinch," in David Cope, *Virtual Music: Computer Synthesis of Musical Style* (Cambridge, Mass.: MIT Press, 2001), pp. 33–82.

104 knight's training . . . Schwarzkopf: See David Shenk, *The Immortal Game* (New York: Doubleday, 2006).

104 "The first time I": Hofstadter, quoted in Bruce Weber, "Mean Chess-Playing Computer Tears at the Meaning of Thought," *New York Times,* February 19, 1996.

104 "article in *Scientific American*": Almost certainly the shocking Feng-hsiung Hsu, Thomas Anantharaman, Murray Campbell, and Andreas Nowatzyk, "A Grandmaster Chess Machine," *Scientific American,* October 1990.

104 See figure "Chess Computer Ratings Over Time," *Scientific American,* October 1990.

105 "To some extent, this match is a defense of the whole human race": Quoted by Hofstadter, "Staring Emmy Straight in the Eye," and attributed to a (since-deleted) 1996 article titled "Kasparov Speaks" at www.ibm.com.

105 "The sanctity of human intelligence": Weber, "Mean Chess-Playing Computer."

105 David Foster Wallace (originally in reference to a tennis match), in "The String Theory," in *Esquire,* July 1996. Collected (under the title "Tennis Player Michael Joyce's Professional Artistry as a Paradigm of Certain Stuff about Choice, Freedom, Discipline, Joy, Grotesquerie, and Human Completeness") in *A Supposedly Fun Thing I'll Never Do Again* (Boston: Little, Brown, 1997).

106 "I personally guarantee": From the press conference after Game 6, as reported by Malcolm Pein of the London Chess Centre.

106 Claude Shannon, "Programming a Computer for Playing Chess," *Philosophical Magazine,* March 1950, the first paper ever written on computer chess.

107 Hofstadter, in Weber, "Mean Chess-Playing Computer."

107 Searle, in ibid.

107 "unrestrained threshold of excellence": Ibid.

108 Deep Blue didn't win it: As Kasparov said at the press conference, "The match was lost by the world champion [and not won by Deep Blue, was the implication] . . . Forget today's game. I mean, Deep Blue hasn't won a single game out of the five." Bewilderingly, he remarked, "It's not yet ready, in my opinion, to win a big contest."

113 checkmate in 262: See Ken Thompson, "The Longest: KRNKNN in 262," *ICGA Journal* 23, no. 1 (2000), pp. 35–36.

113 *"concepts do not always work"*: James Gleick, "Machine Beats Man on Ancient Front," *New York Times,* August 26, 1986.

114 Michael Littman, quoted in Bryn Nelson, "Checkers Computer Becomes Invincible," msnbc.com, July 19, 2007.

114 Garry Kasparov, *How Life Imitates Chess* (New York: Bloomsbury, 2007).

116 Charles Mee, in "Shaped, in Bits, Drips, and Quips," *Los Angeles Times,* October 24, 2004; and in "About the (Re)Making Project," www.charlesmee.org/html/about.html.

116 "doesn't even count": From Kasparov's remarks at the post–Game 6 press conference.

117 Jonathan Schaeffer et al., "Checkers Is Solved," *Science* 317, no. 5844 (September 14, 2007), pp. 1518–22. For more about Chinook, see Jonathan Schaeffer, *One Jump Ahead: Computer Perfection at Checkers* (New York: Springer, 2008).

118 Game 6 commentary available at the IBM website: www.research.ibm.com/deepblue/games/game6/html/comm.txt.

121 Kasparov, *How Life Imitates Chess.*

123 Vin DiCarlo, "Phone and Text Game," at orders.vindicarlo.com/noflakes.

123 "Once you have performed": Mystery, *The Mystery Method: How to Get Beautiful Women into Bed,* with Chris Odom (New York: St. Martin's, 2007).

123 Ted Koppel, in Jack T. Huber and Dean Diggins, *Interviewing America's Top Interviewers: Nineteen Top Interviewers Tell All About What They Do* (New York: Carol, 1991).

124 Schaeffer et al., "Checkers Is Solved."

126 "I decided to opt for unusual openings": Garry Kasparov, "Techmate," *Forbes,* February 22, 1999.

127 Bobby Fischer, interview on Icelandic radio station Útvarp Saga, October 16, 2006.

127 "pushed further and further in": From www.chess960.net.

130 Yasser Seirawan, in his commentary for the Kasparov–Deep Blue rematch, Game 4: www.research.ibm.com/deepblue/games/game4/html/comm.txt.

130 Robert Pirsig, *Zen and the Art of Motorcycle Maintenance* (New York: Morrow, 1974).

130 "Speed Dating with Yaacov and Sue Deyo," interview with Terry Gross, *Fresh Air,* National Public Radio, August 17, 2005. See also Yaacov Deyo and Sue Deyo, *Speed Dating: The Smarter, Faster Way to Lasting Love* (New York: HarperResource, 2002).

6. The Anti-Expert

132 Garry Kasparov, *How Life Imitates Chess* (New York: Bloomsbury, 2007).

133 Jean-Paul Sartre, "Existentialism Is a Humanism," translated by Bernard Frechtman, reprinted (as "Existentialism") in *Existentialism and Human Emotions* (New York: Citadel, 1987).

134 Stephen Jay Gould, *Full House: The Spread of Excellence from Plato to Darwin* (New York: Harmony Books, 1996).

135 René Descartes, *Meditations on First Philosophy.*

135 *The Terminator,* directed by James Cameron (Orion Pictures, 1984).

135 *The Matrix,* directed by Andy Wachowski and Larry Wachowski (Warner Bros., 1999).

136 Douglas R. Hofstadter, *Gödel, Escher, Bach: An Eternal Golden Braid* (New York: Basic Books, 1979).

137 Mark Humphrys, "How My Program Passed the Turing Test," in *Parsing the Turing Test,* edited by Robert Epstein et al. (New York: Springer, 2008).

137 V. S. Ramachandran and Sandra Blakeslee, *Phantoms in the Brain: Probing the Mysteries of the Human Mind* (New York: William Morrow, 1998).

138 Alan Turing, "On Computable Numbers, with an Application to the Entscheidungsproblem," *Proceedings of the London Mathematical Society,* 1937, 2nd ser., 42, no. 1 (1937), pp. 230–65.

139 Ada Lovelace's remarks come from her translation (and notes thereupon) of Luigi Federico Menabrea's "Sketch of the Analytical Engine Invented by Charles Babbage, Esq.," in *Scientific Memoirs,* edited by Richard Taylor (London, 1843).

139 Alan Turing, "Computing Machinery and Intelligence," *Mind* 59, no. 236 (October 1950), pp. 433–60.

139 For more on the idea of "radical choice," see, e.g., Sartre, "Existentialism Is a Humanism," especially Sartre's discussion of a painter wondering "what painting ought he to make" and a student who came to ask Sartre's advice about an ethical dilemma.

140 Aristotle's arguments: See, e.g., *The Nicomachean Ethics.*

140 For a publicly traded company: Nobel Prize winner, and (says the *Economist*) "the most influential economist of the second half of the 20th century," Milton Friedman wrote a piece in the *New York Times Magazine* in 1970 titled "The Social Responsibility of Business Is to Increase Its Profits." The title makes his thesis pretty clear, but Friedman is careful to specify that he means *public* companies: "The situation of the individual proprietor is somewhat different. If he acts to reduce the returns of his enterprise in order to exercise his 'social responsibility' [or in general to do anything whose end is ultimately something other than profit], he is spending his own money, not someone else's . . . That is his right, and I cannot see that there is any objection to his doing so."

140 Ludwig Wittgenstein, *Philosophical Investigations,* translated by G. E. M. Anscombe (Malden, Mass.: Blackwell, 2001).

141 Antonio Machado, "Proverbios y cantares," in *Campos de Castilla* (Madrid: Renacimiento, 1912).

141 Will Wright, quoted in Geoff Keighley, "*Simply* Divine: The Story of Maxis Software," *GameSpot,* www.gamespot.com/features/maxis/index.html.

141 "Unless a man": Bertrand Russell, *The Conquest of Happiness* (New York: Liveright, 1930).

143 Allen Ginsberg, interviewed by Lawrence Grobel, in Grobel's *The Art of the Interview: Lessons from a Master of the Craft* (New York: Three Rivers Press, 2004).

144 Dave Ackley, personal interview.

144 Jay G. Wilpon, "Applications of Voice-Processing Technology in Telecommunications," in *Voice Communication Between Humans and Machines,* edited by David B. Roe and Jay G. Wilpon (Washington, D.C.: National Academy Press, 1994).

145 Timothy Ferriss, *The 4-Hour Workweek: Escape 9–5, Live Anywhere, and Join the New Rich* (New York: Crown, 2007).

146 Stuart Shieber, personal interview. Shieber is the editor of the excellent volume *The Turing Test: Verbal Behavior as the Hallmark of Intelligence* (Cambridge, Mass.: MIT Press, 2004), and his famous criticism of the Loebner Prize is "Lessons from a Restricted Turing Test," *Communications of the Association for Computing Machinery,* April 1993.

147 "The art of general conversation": Russell, *Conquest of Happiness.*

148 Shunryu Suzuki, *Zen Mind, Beginner's Mind* (Boston: Shambhala, 2006).

148 "Commence relaxation": This was from a television ad for Beck's beer. For more information, see Constance L. Hays, "Can Teutonic Qualities Help Beck's Double Its Beer Sales in Six Years?" *New York Times,* November 12, 1998.

148 Bertrand Russell, " 'Useless' Knowledge," in *In Praise of Idleness, and Other Essays* (New York: Norton, 1935); emphasis mine.

148 Aristotle on friendship: In *The Nicomachean Ethics,* specifically books 8 and 9. See also Richard Kraut, "Aristotle's Ethics," in *The Stanford Encyclopedia of Philosophy,* edited by Edward N. Zalta (Summer 2010 ed.). Whereas Plato argues in *The Republic* that "the fairest class [of things is] that which a man who is to be happy [can] love both for its own sake and for the results," Aristotle insists in *The Nicomachean Ethics* that any element of instrumentality in a relationship weakens the quality or nature of that relationship.

149 Philip Jackson, personal interview.

149 *Sherlock Holmes,* directed by Guy Ritchie (Warner Bros., 2009).

7. Barging In

150 Steven Pinker, *The Language Instinct: How the Mind Creates Language* (New York: Morrow, 1994). For more on how listener feedback affects storytelling, see, e.g., Janet B. Bavelas, Linda Coates, and Trudy Johnson, "Listeners as Co-narrators," *Journal of Personality and Social Psychology* 79, no. 6 (2000), 941–52.

150 Bernard Reginster, personal interview. See also Reginster's colleague, philosopher Charles Larmore, who in *The Romantic Legacy* (New York: Columbia University Press, 1996), argues, "We can see the significance of Stendhal's idea [in *Le rouge et le noir*] that the distinctive thing about being natural is that it is *unreflective*." Larmore concludes: "The importance of the Romantic theme of authenticity is that it disabuses us of the idea that life is necessarily better the more [and longer] we think about it."

151 Alan Turing, "Computing Machinery and Intelligence," *Mind* 59, no. 236 (October 1950), pp. 433–60.

152 John Geirland, "Go with the Flow," interview with Mihaly Csikszentmihalyi, *Wired* 4.09 (September 1996).

152 Mihaly Csikszentmihalyi, *Flow: The Psychology of Optimal Experience* (New York: Harper & Row, 1990). See also Mihaly Csikszentmihalyi, *Creativity: Flow and the Psychology of Discovery and Invention* (New York: HarperCollins, 1996); and Mihaly Csikszentmihalyi and Kevin Rathunde, "The Measurement of Flow in Everyday Life: Towards a Theory of Emergent Motivation," in *Developmental Perspectives on Motivation: Nebraska Symposium on Motivation, 1992*, edited by Janis E. Jacobs (Lincoln: University of Nebraska Press, 1993).

154 Dave Ackley, "Life Time," *Dave Ackley's Living Computation*, www.ackleyshack.com/lc/d/ai/time.html.

155 Stephen Wolfram, "A New Kind of Science" (lecture, Brown University, 2003); Stephen Wolfram, *A New Kind of Science* (Champaign, Ill.: Wolfram Media, 2002).

155 Hava Siegelmann, *Neural Networks and Analog Computation: Beyond the Turing Limit* (Boston: Birkhäuser, 1999).

155 Michael Sipser, *Introduction to the Theory of Computation* (Boston: PWS, 1997).

156 Ackley, "Life Time."

156 Noam Chomsky, *Aspects of the Theory of Syntax* (Cambridge, Mass.: MIT Press, 1965).

156 Herbert H. Clark and Jean E. Fox Tree, "Using *Uh* and *Um* in Spontaneous Speaking," *Cognition* 84 (2002), pp. 73–111. See also Jean E. Fox Tree, "Listeners' Uses of *Um* and *Uh* in Speech Comprehension," *Memory & Cognition* 29, no. 2 (2001), pp. 320–26.

158 The first appearance of the word "satisficing" in this sense is Herbert Simon, "Rational Choice and the Structure of the Environment," *Psychological Review* 63 (1956), pp. 129–38.

158 Brian Ferneyhough, quoted in Matthias Kriesberg, "A Music So Demanding That It Sets You Free," *New York Times,* December 8, 2002.

158 Tim Rutherford-Johnson, "Music Since 1960: Ferneyhough: *Cassandra's Dream Song,*" *Rambler,* December 2, 2004, johnsonsrambler.wordpress .com/2004/12/02/music-since-1960-ferneyhough-cassandras-dream -song.

159 "Robert Medeksza Interview—Loebner 2007 Winner," *Ai Dreams,* aidreams.co.uk/forum/index.php?page=67.

160 Kyoko Matsuyama, Kazunori Komatani, Tetsuya Ogata, and Hiroshi G. Okuno, "Enabling a User to Specify an Item at Any Time During System Enumeration: Item Identification for Barge-In-Able Conversational Dialogue Systems," *Proceedings of the International Conference on Spoken Language Processing* (2009).

160 Brian Ferneyhough, in Kriesberg, "Music So Demanding."

161 David Mamet, *Glengarry Glen Ross* (New York: Grove, 1994).

161 For more on back-channel feedback and the (previously neglected) role of the listener in conversation, see, e.g., Bavelas, Coates, and Johnson, "Listeners as Co-narrators."

162 Jack T. Huber and Dean Diggins, *Interviewing America's Top Interviewers: Nineteen Top Interviewers Tell All About What They Do* (New York: Carol, 1991).

163 Clark and Fox Tree, "Using *Uh* and *Um.*"

163 Clive Thompson, "What Is I.B.M.'s Watson?" *New York Times,* June 14, 2010.

163 Nikko Ström and Stephanie Seneff, "Intelligent Barge-In in Conversational Systems," *Proceedings of the International Conference on Spoken Language Processing* (2000).

169 Jonathan Schull, Mike Axelrod, and Larry Quinsland, "Multichat: Persistent, Text-as-You-Type Messaging in a Web Browser for Fluid Multi-person Interaction and Collaboration" (paper presented at the Seventh Annual Workshop and Minitrack on Persistent Conversation, Hawaii International Conference on Systems Science, Kauai, Hawaii, January 2006).

171 Deborah Tannen, *That's Not What I Meant! How Conversational Style Makes or Breaks Relationships* (New York: Ballantine, 1987).

171 For more on the breakdown of strict turn-taking in favor of a more collaborative model of speaking, and its links to everything from intimacy to humor to gender, see, e.g., Jennifer Coates, "Talk in a Play Frame: More on Laughter and Intimacy," *Journal of Pragmatics* 39 (2007), pp. 29–49; and Jennifer Coates, "No Gap, Lots of Overlap: Turn-Taking Patterns in the Talk of Women Friends," in *Researching Language and Literacy in Social Context,* edited by David Graddol, Janet Maybin, and Barry Stierer (Philadelphia: Multilingual Matters, 1994), pp. 177–92.

8. *The World's Worst Deponent*

174 Albert Mehrabian, *Silent Messages* (Belmont, Calif.: Wadsworth, 1971).

175 For more on telling stories backward, see, e.g., Tiffany McCormack, Alexandria Ashkar, Ashley Hunt, Evelyn Chang, Gent Silberkleit, and R. Edward Geiselman, "Indicators of Deception in an Oral Narrative: Which Are More Reliable?" *American Journal of Forensic Psychology* 30, no. 4 (2009), pp. 49–56.

175 For more on objections to form, see, e.g., Paul Bergman and Albert Moore, *Nolo's Deposition Handbook* (Berkeley, Calif.: Nolo, 2007). For additional research on lie detection in the realm of electronic text, see, e.g., Lina Zhou, "An Empirical Investigation of Deception Behavior in Instant Messaging," *IEEE Transactions on Professional Communication* 48, no. 2 (2005), pp. 147–60.

176 "unasking" of the question: This phrasing comes from both Douglas R. Hofstadter, *Gödel, Escher, Bach: An Eternal Golden Braid* (New York: Basic Books, 1979), and Robert Pirsig, *Zen and the Art of Motorcycle Maintenance* (New York: Morrow, 1974). Pirsig also describes

mu using the metaphor of a digital circuit's "high impedance" (a.k.a. "floating ground") state: neither 0 nor 1.

177 Eben Harrell, "Magnus Carlsen: The 19-Year-Old King of Chess," *Time,* December 25, 2009.

178 Lawrence Grobel, *The Art of the Interview: Lessons from a Master of the Craft* (New York: Three Rivers Press, 2004).

178 For more on the topic of our culture's rhetorical "minimax" attitude, see, e.g., Deborah Tannen, *The Argument Culture* (New York: Random House, 1998).

180 Paul Ekman, *Telling Lies: Clues to Deceit in the Marketplace, Politics, and Marriage* (New York: Norton, 2001).

181 Leil Lowndes, *How to Talk to Anyone* (London: Thorsons, 1999).

181 Neil Strauss, *The Game: Penetrating the Secret Society of Pickup Artists* (New York: ReganBooks, 2005).

181 Larry King, *How to Talk to Anyone, Anytime, Anywhere* (New York: Crown, 1994).

181 Dale Carnegie, *How to Win Friends and Influence People* (New York: Pocket, 1998).

185 David Foster Wallace, *Infinite Jest* (Boston: Little, Brown, 1996).

188 Melissa Prober, personal interview.

188 Mike Martinez, personal interview.

190 David Sheff, personal interview.

191 Ekman, *Telling Lies.*

192 Will Pavia tells his story of being fooled in the 2008 Loebner Prize competition in "Machine Takes on Man at Mass Turing Test," *Times* (London), October 13, 2008.

194 Dave Ackley, personal interview.

9. *Not Staying Intact*

196 Bertrand Russell, *The Conquest of Happiness* (New York: Liveright, 1930).

196 Racter, *The Policeman's Beard Is Half Constructed* (New York: Warner Books, 1984).

198 David Levy, Roberta Catizone, Bobby Batacharia, Alex Krotov, and Yorick Wilks, "CONVERSE: A Conversational Companion," *Proceedings of the First International Workshop of Human-Computer Conversation* (Bellagio, Italy, 1997).

198 Yorick Wilks, "On Whose Shoulders?" (Association for Computational Linguistics Lifetime Achievement Award speech, 2008).

199 Thomas Whalen, "Thom's Participation in the Loebner Competition 1995; or, How I Lost the Contest and Re-evaluated Humanity," thomwhalen.com-ThomLoebner1995.html.

200 The PARRY and ELIZA transcript comes from their encounter on September 18, 1972.

201 Michael Gazzaniga, *Human: The Science Behind What Makes Us Unique* (New York: Ecco, 2008).

201 Mystery, *The Mystery Method: How to Get Beautiful Women into Bed*, with Chris Odom (New York: St. Martin's, 2007).

201 Ross Jeffries, in "Hypnotists," *Louis Theroux's Weird Weekends*, BBC Two, September 25, 2000.

202 Richard Bandler and John Grinder, *Frogs into Princes: Neuro Linguistic Programming* (Moab, Utah: Real People Press, 1979).

202 Will Dana, in Lawrence Grobel, *The Art of the Interview: Lessons from a Master of the Craft* (New York: Three Rivers Press, 2004).

202 David Sheff, personal interview.

203 Racter, *Policeman's Beard.*

204 Ludwig Wittgenstein, *Philosophical Investigations*, translated by G. E. M. Anscombe (Malden, Mass.: Blackwell, 2001).

204 My money—and that of many others: See also the famous 1993 accusation by early blogger (in fact, inventor of the term "weblog") Jorn Barger: " 'The Policeman's Beard' Was Largely Prefab!" www.robot wisdom.com/ai/racterfaq.html.

204 a YouTube video: This particular video is of the bot Cassandra, in development by the ejTalk corporation. http://www.youtube.com/watch?v= 0Tqt2TurCnI.

205 Salvador Dalí, "Preface: Chess, It's Me," translated by Albert Field, in Pierre Cabanne, *Dialogues with Marcel Duchamp* (Cambridge, Mass.: Da Capo, 1987).

205 Richard S. Wallace, "The Anatomy of A.L.I.C.E.," in *Parsing the Turing Test*, edited by Robert Epstein et al. (New York: Springer, 2008).

205 Hava Siegelmann, personal interview.

205 George Orwell, "Politics and the English Language," *Horizon* 13, no. 76 (April 1946), pp. 252–65.

206 Roger Levy, personal interview.

206 Dave Ackley, personal interview.

206 *Freakonomics* (Levitt and Dubner, see below) notes that "the Greater Dallas Council on Alcohol and Drug Abuse has compiled an extraordinarily entertaining index of cocaine street names."

207 Harold Bloom, *The Anxiety of Influence: A Theory of Poetry* (New York: Oxford University Press, 1973).

207 Ezra Pound's famous battle cry of modernism, "Make it new," comes from his translation of the Confucian text *The Great Digest,* a.k.a. *The Great Learning.*

207 Garry Kasparov, *How Life Imitates Chess* (New York: Bloomsbury, 2007).

207 Sun Tzu, *The Art of War,* translated by John Minford (New York: Penguin, 2003).

208 The phrase "euphemism treadmill" comes from Steven Pinker, *The Blank Slate* (New York: Viking, 2002). See also W. V. Quine, "Euphemism," in *Quiddities: An Intermittently Philosophical Dictionary* (Cambridge, Mass.: Belknap, 1987).

208 The controversy over Rahm Emanuel's remark appears to have originated with Peter Wallsten, "Chief of Staff Draws Fire from Left as Obama Falters," *Wall Street Journal,* January 26, 2010.

208 Rosa's Law, S.2781, 2010.

208 "Mr. Burton's staff": Don Van Natta Jr., "Panel Chief Refuses Apology to Clinton," *New York Times,* April 23, 1998.

209 Will Shortz, quoted in Jesse Sheidlower, "The Dirty Word in 43 Down," *Slate Magazine,* April 6, 2006.

209 Steven D. Levitt and Stephen J. Dubner, *Freakonomics: A Rogue Economist Explores the Hidden Side of Everything* (New York: William Morrow, 2005).

209 Guy Deutscher, *The Unfolding of Language: An Evolutionary Tour of Mankind's Greatest Invention* (New York: Metropolitan Books, 2005).

210 Joseph Weizenbaum, *Computer Power and Human Reason: From Judgment to Calculation* (San Francisco: W. H. Freeman, 1976).

210 The effect that photons have on the electrons they are measuring is called the Compton effect; the paper where Heisenberg uses this to lay the foundation for his famous "uncertainty principle" is "Über den anschaulichen Inhalt der quantentheoretischen Kinematik und Mechanik," *Zeitschrift für Physik* 43 (1927), pp. 172–98, available

in English in *Quantum Theory and Measurement,* edited by John Archibald Wheeler and Wojciech Hubert Zurek (Princeton, N.J.: Princeton University Press, 1983).

210 Deborah Tannen's *That's Not What I Meant! How Conversational Style Makes or Breaks Relationships* (New York: Ballantine, 1987) has illuminating sample dialogues of how trying to ask a question "neutrally" can go horribly wrong.

210 A famous study on wording and memory, and the one from which the car crash language is taken, is from Elizabeth F. Loftus and John C. Palmer, "Reconstruction of Automobile Destruction: An Example of the Interaction Between Language and Memory," *Journal of Verbal Learning and Verbal Behavior* 13, no. 5 (October 1974), pp. 585–89.

211 For more on the "or not" wording, see, e.g., Jon Krosnick, Eric Shaeffer, Gary Langer, and Daniel Merkle, "A Comparison of Minimally Balanced and Fully Balanced Forced Choice Items" (paper presented at the annual meeting of the American Association for Public Opinion Research, Nashville, August 16, 2003).

211 For more on how asking about one dimension of life can (temporarily) alter someone's perception of the rest of their life, see Fritz Strack, Leonard Martin, and Norbert Schwarz, "Priming and Communication: Social Determinants of Information Use in Judgments of Life Satisfaction," *European Journal of Social Psychology* 18, no. 5 (1988), pp. 429–42. Broadly, this is referred to as a type of "focusing illusion."

211 Robert Creeley and Archie Rand, *Drawn & Quartered* (New York: Granary Books, 2001).

211 Marcel Duchamp, *Nude Descending a Staircase, No. 2* (1912), Philadelphia Museum of Art.

212 Hava Siegelmann, *Neural Networks and Analog Computation: Beyond the Turing Limit* (Boston: Birkhäuser, 1999).

212 Ackley, personal interview.

213 Plato, *Symposium,* translated by Benjamin Jowett, in *The Dialogues of Plato, Volume One* (New York: Oxford University Press, 1892).

213 Phil Collins, "Two Hearts," from *Buster: The Original Motion Picture Soundtrack.*

213 John Cameron Mitchell and Stephen Trask, *Hedwig and the Angry Inch,* directed by John Cameron Mitchell (Killer Films, 2001).

214 Spice Girls, "2 Become 1," *Spice* (Virgin, 1996).

214 *Milk,* directed by Gus Van Sant (Focus Features), 2008.

215 Kevin Warwick, personal interview.

215 Thomas Nagel, "What Is It Like to Be a Bat?" *Philosophical Review* 83, no. 4 (October 1974), pp. 435–50.

216 Douglas R. Hofstadter, *I Am a Strange Loop* (New York: Basic Books, 2007).

216 Gazzaniga, *Human.*

217 Russell, *Conquest of Happiness.*

217 Roberto Caminiti, Hassan Ghaziri, Ralf Galuske, Patrick Hof, and Giorgio Innocenti, "Evolution Amplified Processing with Temporally Dispersed Slow Neuronal Connectivity in Primates," *Proceedings of the National Academy of Sciences* 106, no. 46 (November 17, 2009), pp. 19551–56.

218 The Bach cantata is 197, "Gott ist unsre Zuversicht." For more, see Hofstadter's *I Am a Strange Loop.*

218 Benjamin Seider, Gilad Hirschberger, Kristin Nelson, and Robert Levenson, "We Can Work It Out: Age Differences in Relational Pronouns, Physiology, and Behavior in Marital Conflict," *Psychology and Aging* 24, no. 3 (September 2009), pp. 604–13.

10. High Surprisal

222 Claude Shannon, "A Mathematical Theory of Communication," *Bell System Technical Journal* 27 (1948), pp. 379–423, 623–56.

223 average American teenager: Katie Hafner, "Texting May Be Taking a Toll," *New York Times,* May 25, 2009.

226 The two are in fact related: For more information on the connections between Shannon (information) entropy and thermodynamic entropy, see, e.g., Edwin Jaynes, "Information Theory and Statistical Mechanics," *Physical Review* 106, no. 4, (May 1957), pp. 620–30; and Edwin Jaynes, "Information Theory and Statistical Mechanics II," *Physical Review* 108, no. 2 (October 1957), pp. 171–90.

228 Donald Barthelme, "Not-Knowing," in *Not-Knowing: The Essays and Interviews of Donald Barthelme,* edited by Kim Herzinger (New York: Random House, 1997).

229 Jonathan Safran Foer, *Extremely Loud and Incredibly Close* (Boston: Houghton Mifflin, 2005).

229 The cloze test comes originally from W. Taylor, "Cloze procedure: A New Tool for Measuring Readability," *Journalism Quarterly* 30 (1953), pp. 415–33.

231 Mystery, *The Mystery Method: How to Get Beautiful Women into Bed,* with Chris Odom (New York: St. Martin's, 2007).

232 Scott McDonald and Richard Shillcock, "Eye Movements Reveal the On-Line Computation of Lexical Probabilities During Reading," *Psychological Science* 14, no. 6, (November 2003), pp. 648–52.

232 Keith Rayner, Katherine Binder, Jane Ashby, and Alexander Pollatsek, "Eye Movement Control in Reading: Word Predictability Has Little Influence on Initial Landing Positions *in* Words" (emphasis mine, as they reference its effects *on* words). *Vision Research* 41, no. 7 (March 2001), pp. 943–54. For more on entropy's effect on reading, see Keith Rayner, "Eye Movements in Reading and Information Processing: 20 Years of Research," *Psychological Bulletin* 124, No. 3, (November 1998), pp. 372–422; Steven Frisson, Keith Rayner, and Martin J. Pickering, "Effects of Contextual Predictability and Transitional Probability on Eye Movements During Reading," *Journal of Experimental Psychology: Learning, Memory, and Cognition* 31, No. 5 (September 2005), pp. 862–77; Reinhold Kliegl, Ellen Grabner, Martin Rolfs, and Ralf Engbert, "Length, Frequency, and Predictability Effects of Words on Eye Movements in Reading," *European Journal of Cognitive Psychology* 16, nos. 1–2 (January–March 2004), pp. 262–84.

233 Laurent Itti and Pierre Baldi, "Bayesian Surprise Attracts Human Attention," *Vision Research* 49, no. 10 (May 2009), pp. 1295–306. See also, Pierre Baldi and Laurent Itti, "Of Bits and Wows: A Bayesian Theory of Surprise with Applications to Attention," *Neural Networks* 23, no. 5 (June 2010), pp. 649–66; Linda Geddes, "Model of Surprise Has 'Wow' Factor Built In," *New Scientist,* January 2009; Emma Byrne, "Surprise Moves Eyes," Primary Visual Cortex, October 2008; T. Nathan Mundhenk, Wolfgang Einhäuser, and Laurent Itti, "Automatic Computation of an Image's Statistical Surprise Predicts Performance of Human Observers on a Natural Image Detection Task," *Vision Research* 49, no. 13 (June 2009), pp. 1620–37.

233 al Qaeda videos: In Kim Zetter, "Researcher's Analysis of al Qaeda Images Reveals Surprises," *Wired,* August 2, 2007.

233 fashion industry: Neal Krawetz, "Body by Victoria," *Secure Computing*

blog, www.hackerfactor.com/blog/index.php?/archives/322-Body-By
-Victoria.html.

236 T. S. Eliot, "The Love Song of J. Alfred Prufrock," *Poetry*, June 1915.

237 Marcel Duchamp, *Fountain* (1917).

238 C .D. Wright, "Tours," in *Steal Away* (Port Townsend, Wash.: Copper Canyon Press, 2002).

238 Milan Kundera, *The Unbearable Lightness of Being* (New York: Harper & Row, 1984).

238 David Shields, quoted in Bond Huberman, "I Could Go On Like This Forever," *City Arts*, July 1, 2008.

238 Roger Ebert, review of *Quantum of Solace*, November 12, 2008, at rogerebert.suntimes.com.

239 Matt Mahoney, "Text Compression as a Test for Artificial Intelligence," *Proceedings of the Sixteenth National Conference on Artificial Intelligence and the Eleventh Innovative Applications of Artificial Intelligence Conference* (Menlo Park, Calif.: American Association for Artificial Intelligence, 1999). See also Matt Mahoney, *Data Compression Explained* (San Jose, Calif.: Ocarina Networks, 2010), www.mattmahoney.net/dc/dce.html.

239 Annie Dillard, *An American Childhood* (New York: Harper & Row, 1987).

240 Eric Hayot, in "Somewhere Out There," episode 374 of *This American Life*, February 13, 2009.

240 *Three Colors: White*, directed by Krzysztof Kieślowski (Miramax, 1994).

240 David Bellos, "I, Translator," *New York Times*, March 20, 2010.

241 Douglas R. Hofstadter, *Gödel, Escher, Bach: An Eternal Golden Braid* (New York: Basic Books, 1979).

242 "Six Years Later: The Children of September 11," *The Oprah Winfrey Show*, September 11, 2007.

242 George Bonanno, "Loss, Trauma, and Human Resilience: Have We Underestimated the Human Capacity to Thrive After Extremely Adverse Events?" *American Psychologist* 59, no. 1 (January 2004), pp. 20–28. See also George Bonanno, *The Other Side of Sadness: What the New Science of Bereavement Tells Us About Life After Loss* (New York: Basic Books, 2009).

243 Robert Pirsig, *Zen and the Art of Motorcycle Maintenance* (New York: Morrow, 1974).

245 It's widely held: See, e.g., papers by the University of Edinburgh's Sharon Goldwater, Brown University's Mark Johnson, UC Berkeley's Thomas Griffiths, the University of Wisconsin's Jenny Saffran, the Moss Rehabilitation Research Institute's Dan Mirman, and the University of Pennsylvania's Daniel Swingley, among others.

246 Eugene Charniak, personal interview.

246 Shannon, "Mathematical Theory of Communication."

246 *The American Heritage Book of English Usage: A Practical and Authoritative Guide to Contemporary English* (Boston: Houghton Mifflin, 1996).

247 "attorney general": These three examples taken from Bill Bryson, *The Mother Tongue: English and How It Got That Way* (New York: Morrow, 1990).

247 Dave Matthews Band, "You and Me," *Big Whiskey and the GrooGrux King* (RCA, 2009).

248 Norton Juster, *The Phantom Tollbooth* (New York: Epstein & Carroll, 1961).

248 Guy Blelloch, "Introduction to Data Compression," manuscript, 2001.

249 David Foster Wallace, "Authority and American Usage," in *Consider the Lobster* (New York: Little, Brown, 2005).

251 Takeshi Murata, "Monster Movie" (2005).

251 Kanye West, "Welcome to Heartbreak," directed by Nabil Elderkin (2009).

252 Kundera, *Unbearable Lightness of Being.*

253 Hofstadter, *Gödel, Escher, Bach.*

256 Kundera, *Unbearable Lightness of Being.*

257 Timothy Ferriss, interview with Leon Ho, *Stepcase Lifehack,* June 1, 2007.

257 Heather McHugh, "In Ten Senses: Some Sentences About Art's Senses and Intents" (lecture, University of Washington, Solomon Katz Distinguished Lectures in the Humanities, December 4, 2003).

257 Forrest Gander, *As a Friend* (New York: New Directions, 2008).

258 Claude Shannon, "Prediction and Entropy of Printed English," *Bell System Technical Journal* 30, no. 1 (1951), pp. 50–64.

259 Shannon, "Mathematical Theory of Communication."

11. Conclusion: The Most Human Human

260 David Levy, *Love and Sex with Robots* (New York: HarperCollins, 2007).

262 Robert Epstein, "My Date with a Robot," *Scientific American Mind,* June/July 2006.

262 Garry Kasparov, *How Life Imitates Chess* (New York: Bloomsbury, 2007).

263 Ray Kurzweil, *The Singularity Is Near: When Humans Transcend Biology* (New York: Viking, 2005).

264 bacteria rule the earth: See Stephen Jay Gould, *Full House: The Spread of Excellence from Plato to Darwin* (New York: Harmony Books 1996).

Epilogue: The Unsung Beauty of the Glassware Cabinet

267 The idea of the Cornell box dates back to Cindy M. Goral, Kenneth E. Torrance, Donald P. Greenberg, and Bennett Battaile, "Modeling the Interaction of Light Between Diffuse Surfaces," *Computer Graphics* (*SIGGRAPH Proceedings*) 18, no. 3 (July 1984), pp. 213–22.

267 Devon Penney, personal interview.

269 Eduardo Hurtado, "Instrucciones para pintar el cielo" ("How to Paint the Sky"), translated by Mónica de la Torre, in *Connecting Lines: New Poetry from Mexico,* edited by Luis Cortés Bargalló and Forrest Gander (Louisville, Ky.: Sarabande Books, 2006).

270 Bertrand Russell, "In Praise of Idleness," in *In Praise of Idleness, and Other Essays* (New York: Norton, 1935).